I0010180

Deep Learning for Genomics

Data-driven approaches for genomics applications in life sciences and biotechnology

Upendra Kumar Devisetty

BIRMINGHAM—MUMBAI

Deep Learning for Genomics

Copyright © 2022 Packt Publishing

All rights reserved. No part of this book may be reproduced, stored in a retrieval system, or transmitted in any form or by any means, without the prior written permission of the publisher, except in the case of brief quotations embedded in critical articles or reviews.

Every effort has been made in the preparation of this book to ensure the accuracy of the information presented. However, the information contained in this book is sold without warranty, either express or implied. Neither the author(s), nor Packt Publishing or its dealers and distributors, will be held liable for any damages caused or alleged to have been caused directly or indirectly by this book.

Packt Publishing has endeavored to provide trademark information about all of the companies and products mentioned in this book by the appropriate use of capitals. However, Packt Publishing cannot guarantee the accuracy of this information.

Publishing Product Manager: Dhruv Jagdish Kataria

Content Development Editor: Priyanka Soam

Technical Editor: Rahul Limbachiya

Copy Editor: Safis Editing

Project Coordinator: Farheen Fathima

Proofreader: Safis Editing

Indexer: Rekha Nair

Production Designer: Mohamed Huzair

Marketing Coordinators: Shifa Ansari, Abeer Riyaz Dawe

First published: October 2022

Production reference: 1311022

Published by Packt Publishing Ltd.

Livery Place

35 Livery Street

Birmingham

B3 2PB, UK.

ISBN 978-1-80461-544-7

www.packt.com

Contributors

About the author

Upendra Kumar Devisetty has a Ph.D. in agriculture and over 12 years of experience working in Next-Generation Sequencing. He has a deep background in genomics and bioinformatics with a specialization in applying predictive analytics across a varied set of genomics problems in life sciences. Dr. Devisetty is currently working as a senior data science manager at Greenlight Biosciences, where he leads a team of bioinformatics scientists and data scientists to support the various bioinformatics and data science projects at Greenlight Biosciences with a mission to create mRNA-based solutions that can provide a cleaner environment and healthier people.

About the reviewer

Urminder Singh is a computer scientist and bioinformatician. His diverse research interests include understanding novel gene evolution, cancer genomics, machine learning in medicine, sociogenomics, and algorithms for big heterogeneous data. You can find him online at `urmi-21.github.io`.

Table of Contents

Preface   xi

Part 1 – Machine Learning in Genomics

1

Introducing Machine Learning for Genomics   3

What is machine learning?	4	Python programming language	6
Why machine learning for genomics?	5	Visualization	7
Machine learning for genomics in life sciences and biotechnology	6	Biopython	7
		Scikit-learn	7
Exploring machine learning software	6	**Summary**	7

2

Genomics Data Analysis   9

Technical requirements	10	Introduction to Biopython for genomic data analysis	18
Installing Biopython	10		
Matplotlib	12	What is Biopython?	18
What is a genome?	12	Genomic data analysis use case – Sequence analysis of Covid-19	21
Genome sequencing	13	Calculating GC content	24
Sanger sequencing of nucleic acids	13	Calculating nucleotide content	24
Evolution of next-generation sequencing	14	Dinucleotide content	25
Analysis of genomic data	14	Modeling	27
Steps in genomics data analysis	15	Motif finder	29
		Summary	30

3

Machine Learning Methods for Genomic Applications   31

Technical requirements	32	**An ML use case for genomics –**	
Python packages	32	**Disease prediction**	**40**
ML libraries	32	Data collection	41
		Data preprocessing	42
Genomics big data	**33**	EDA	43
Supervised and unsupervised ML	**34**	Data transformation	45
Supervised ML	35	Data splitting	47
Unsupervised ML	37	Model training	48
		Model evaluation	48
ML for genomics	**38**	**ML challenges in genomics**	**51**
The basic workflow of ML in genomics	38	**Summary**	**52**

Part 2 – Deep Learning for Genomic Applications

4

Deep Learning for Genomics   55

Understanding what deep learning is		Gene regulatory networks	77
and how it works	**56**	Single-cell RNA sequencing	77
Neural network definition	56	**Introducing deep learning**	
Anatomy of deep neural networks	**56**	**algorithms and Python libraries**	**78**
Key concepts of DNNs	59	General deep learning libraries	78
An example of how neural networks work	66	Deep learning libraries for genomics	79
DNN architectures	69	**Summary**	**80**
DNNs for genomics	**74**		
Deep learning workflow for genomics	74		
Broad application of DNNs in genomics	76		
Protein structure predictions	77		
Regulatory genomics	77		

5

Introducing Convolutional Neural Networks for Genomics 81

Introduction to CNNs	82	DeepBind	90
What are CNNs?	83	DeepInsight	91
Transfer Learning	88	DeepChrome	92
		DeepVariant	92
CNNs for genomics	89	**Summary**	**93**
Applications of CNNs in genomics	90		

6

Recurrent Neural Networks in Genomics 95

What are RNNs?	96	**Applications and use cases of RNNs in genomics**	**106**
Introducing RNNs	98	DeepNano	106
How do RNNs work?	99	ProLanGo	107
Different RNN architectures	102	DanQ	108
Bidirectional RNNs (BiLSTM)	103	Understanding RNNs through Transcription Factor Binding Site (TFBS) predictions	109
LSTMs and GRUs	103		
Different types of RNNs	105	**Summary**	**115**

7

Unsupervised Deep Learning with Autoencoders 117

What is unsupervised DL?	118	Types of autoencoders	127
Types of unsupervised DL	118	**Autoencoders for genomics**	**130**
Clustering	118	Gene expression	130
Anomaly detection	119	Use case – Predicting gene expression from TCGA pan-cancer RNA-Seq data using denoising autoencoders	131
Association	121		
What are autoencoders?	121		
Properties of autoencoders	122	**Summary**	**136**
How do autoencoders work?	122		
Architecture of autoencoders	124		

8

GANs for Improving Models in Genomics   137

What are GANs?   138

Differences between Discriminative and
Generative models   138
Intuition about GANs   139
How do GANs work?   140

**Challenges working with genomics
datasets**   143
What is synthetic data?   144

How can GANs help improve models? 146

**Practical applications of GANs in
genomics**   147
Analysis of ScRNA-Seq data   148
Generation of DNA   149
Using GANs for augmenting population-scale
genomics data   150

Summary   150

Part 3 – Operationalizing models

9

Building and Tuning Deep Learning Models   153

Technical requirements   154

DL life cycle   154

Data processing   157
Data collection   157
Data wrangling   158
Feature engineering   159

Developing models   160
Selecting an appropriate algorithm   161
Model training   161

Tuning the models   163
Hyperparameter tuning   164
Hyperparameter tuning libraries   166

Classification metrics or performance statistics 169
Visualizing performance   172
Regression metrics   173

**Use case – Predicting the binding site
location of the JunD TF**   174
Framing the TFBS prediction problem in
terms of DL   175
Processing the data   175
Model training   179

Summary   181

10

Model Interpretability in Genomics   183

What is model interpretability?	**184**	Global surrogate	192
Black-box model interpretability	185	LIME	193
		Shapley value	193
Unlocking business value from		ExSum	194
model interpretability	**187**	Saliency map	194
Better business decisions	187	**Use case – Model interpretability for**	
Building trust	189	**genomics**	**194**
Profitability	189	Data collection	195
		Feature extraction	195
Model interpretability methods in		Target labels	196
genomics	**189**	Train-test split	196
Partial dependence plot	190	Creating a CNN architecture	197
Individual conditional expectation	191	**Summary**	**201**
Permuted feature importance	191		

11

Model Deployment and Monitoring   203

Technical requirements	**204**	**Monitoring models using advanced**	
Streamlit	204	**tools**	**220**
Hugging Face	205	Why monitor models?	220
		Reasons for model degradation	220
Introducing model deployment	**205**	How to monitor DL models	222
Steps in model deployment	206	Advanced tools for model monitoring	222
Types of model deployment	208	Addressing drifts	223
Deploying models as services	209	**Summary**	**223**
A use case for deploying a DL model as a web			
service – building a Streamlit application of			
the CNN model	211		

12

Challenges, Pitfalls, and Best Practices for Deep Learning in Genomics   225

Deep learning challenges regarding
genomics   226
Lack of flexible tools   226
Fewer biological samples   226
Computational resource requirements   227
Expertise in DL frameworks   227
Lack of high-quality labeled data   227
Lack of model interpretability   227

Common pitfalls for applying deep
learning to genomics   228
Confounding   228
Data leakage   228
Imbalanced data   229
Improper model comparisons   230

Best practices for applying deep

learning to genomics   230
Understand the problem and know your data
better   230
A simple model for a simple problem   231
Establish a baseline for your model   231
Ensure reproducibility   232
Using pre-existing models for genomics   232
Do not reinvent the rule   232
Tune hyperparameters automatically   232
Focus on feature engineering   233
Normalize the data   233
Always perform model interpretation   233
Avoid overfitting   234

Summary   234

Index   237

Other Books You May Enjoy   248

Preface

Deep learning is the subset of machine learning based on artificial neural networks with representative learning using vast amounts of data. Machine learning is a subcomponent of artificial intelligence, which includes sophisticated algorithms that enable machines to mimic human intelligence to perform human tasks automatically. Both deep learning and machine learning help automatically detect meaningful patterns from data without explicit programming. Machine learning and deep learning have completely changed the way that we live these days. We rely on these so much that it's hard to imagine a day without using any of these in some way or another, whether it is via the spam filtering of emails, product recommendations, or speech recognition. Both machine learning and specifically deep learning have been adopted by the scientific community in areas such as biology, genomics, bioinformatics, and computational biology. **High-throughput technologies (HTS)** such as **next-generation sequencing (NGS)** have made a significant contribution to genomics to study complex biological phenomena at a single-base-pair resolution on an unprecedented scale, facilitating an era of big data genomics. To get meaningful and novel biological insights from this big data, most of the algorithms are currently based on machine learning and, lately, deep learning methodologies to provide higher levels of accuracy in specific tasks related to genomics than state-of-the-art rule-based algorithms. Given the growing trend in the perception and application of machine learning and deep learning in genomics, research professionals, scientists, and managers require a good understanding of this exciting field to equip them with the necessary tools, technologies, and general guidelines to assist them in the selection of machine learning and deep learning methods for handling genomics data and accelerating data-driven decision-making in industries related to life sciences and biotechnology.

Throughout this book, we will learn how to apply deep learning approaches to solve real-world problems in genomics, interpret biological insights from deep learning models built from genomic datasets, and finally, operationalize deep learning models using open source tools to enable predictions for end users.

Who is this book for?

This book aims to practically introduce machine learning and deep learning for genomic applications that can transform genomics data into novel biological insights. It provides both the theoretical fundamentals and hands-on sections to give a taste of how machine learning and deep learning can be leveraged in real-world applications in the life sciences and biotech industries. This book covers a range of topics that are not currently available in other textbooks. The book also includes the challenges, pitfalls, and best practices when applying machine learning and deep learning to real-world scenarios. Each chapter of the book has code written in Python with industry-standard machine learning and deep learning libraries and frameworks such as Keras that the audience can reproduce in their working environment. This book is designed to cater to the needs of researchers, bioinformaticians, and data scientists in both academia and industry who want to leverage machine learning and deep learning technologies in genomic applications to extract insights from sets of big data. Managers and leaders who are already established in the life sciences and biotechnology sectors will not only find this book useful but can also adopt these methodologies to identify patterns, come up with predictions, and thereby contribute to data-driven decision-making in their respective companies.

The book is divided into three different parts. The first part introduces the fundamentals of genomic data analysis and machine learning. In this part, we will introduce the basic concept of genomic data analysis and discuss what machine learning is and why it is important for genomics and what value machine learning will bring to the life sciences and biotechnology industries. The second part will transition the readers from machine learning to deep learning and introduce them to the basic concepts of deep learning and diverse deep learning algorithms, using real-world examples to transform raw genomics data into biological insights. The final part will describe how to operationalize deep learning models using open source tools to enable predictions for end users. In this part, you will learn how to build and tune state-of-the-art machine learning models using Python and industry-standard libraries to derive biological insights from large amounts of multimodal genomic datasets and how to deploy these models on several cloud platforms such as AWS and Azure. The last chapter in the final part is fully dedicated to the current challenges for deep learning approaches to genomics and the potential pitfalls and how to avoid them using best practices.

What this book covers

Chapter 1, *Introducing Machine Learning for Genomics*, provides a brief history of the field of genomics and the practical application of machine learning methods to genomics, in addition to some of the technologies that this book will use.

Chapter 2, *Genomics Data Analysis*, gives readers a quick primer on data analysis in genomics. Using the Python programming language, readers will be able to make sense of the vast amounts of genomics data available and extract biological insights.

Chapter 3, Machine Learning Methods for Genomic Applications, introduces the reader to the two most important machine learning methods (supervised and unsupervised) and some of the important elements of standard machine learning pipelines. It also includes the practical real-world applications of supervised and unsupervised algorithms for genomics data analysis in the life sciences and biotechnology industries.

Chapter 4, Deep Learning for Genomics, will teach the reader about the fundamental concepts of deep learning, different types of deep learning models, and different deep learning Python libraries.

Chapter 5, Introducing Convolutional Neural Networks for Genomics, gives the reader a taste of **Convolutional Neural Networks** (**CNNs**), a type of deep neural network that is primarily used for sequence data, and shows how CNNs have superior performance compared to other deep learning methods.

Chapter 6, Recurrent Neural Networks in Genomics, introduces reinforcement learning techniques such as **Recurrent Neural Networks** (**RNNs**) and LSTMs and shows how they are currently being applied in several applications.

Chapter 7, Unsupervised Deep Learning with Autoencoders, introduces unsupervised deep learning, different methods of unsupervised deep learning, specifically Autoencoders, and its application in genomics.

Chapter 8, GANs for Improving Models in Genomics, introduces **Generative Adversarial Networks** (**GANs**) and how they can be used to improve deep neural networks trained on genomics datasets for predictive modeling.

Chapter 9, Building and Tuning Deep Learning Models, describes how to build and tune machine learning and deep learning models and deploy the final models across various computational systems and several platforms.

Chapter 10, Model Interpretability in Genomics, introduces the reader to how to interpret machine learning and deep learning models. The model interpretability introduced here helps readers to understand a model's decision and why businesses are interested in model interpretability for creating trust, gaining profitability, and so on.

Chapter 11, Model Deployment and Monitoring, teaches the reader how to take the model they built on Google Colab and deploy it for predictions using open source tools such as Streamlit and Hugging Face. In addition, this chapter also describes how to monitor models using advanced tools and how monitoring is a key metric for businesses.

Chapter 12, Challenges, Pitfalls, and Best Practices for Deep Learning in Genomics, informs the reader of the challenges and pitfalls associated with applying machine learning and deep learning methodologies to genomics applications. It also covers the best practices for building end-to-end machine learning and deep learning models and applying them to genomic datasets.

To get the most out of this book

The book aims to keep it self-contained as possible. To extract the maximum value out of this book, a basic to intermediate knowledge of Python programming is recommended and a background in genomics, statistics, and bioinformatics and some knowledge of data science is a must. In addition, readers are expected to know the basics of machine learning and associated machine learning algorithms, such as regression and classification. The book provides a hands-on approach to implementation and associated deep learning methodologies that will have you up-and-running and productive in no time. At the end of the book, you will be able to put your knowledge to work with this practical guide.

Software covered in the book	Operating system requirements
Python 3	Windows, macOS, or Linux
Juptyer Notebook	Windows, macOS, or Linux
Streamlit	Any modern-day web browser
Google Colab	Windows, macOS, or Linux
Hugging Face	Any modern-day web browser
Keras	Windows, macOS, or Linux

If you are using the digital version of this book, we advise you to type the code yourself or access the code from the book's GitHub repository. This will ensure you avoid any potential error related to copying and pasting of code.

Download the example code files

You can download the example code files for this book from GitHub at `https://github.com/PacktPublishing/Deep-Learning-for-Genomics-`. Any updates to the code will be reflected in the GitHub repository. We also have other code bundles from our rich catalog of books and videos available at: `https://github.com/PacktPublishing/»https://github.com/PacktPublishing/`. Check them out!

Conventions used

There are several text conventions used throughout this book.

Code in text: Indicates code words in the text, folder names, filenames, file extensions, pathnames, dummy URLs, user input, and Twitter handles. Here is an example: "Mount the downloaded `WebStorm-10*.dmg` disk image file as another disk in your system."

A block of code is set as follows:

```
# covid19_features.py
from Bio import SeqIO
```

When we wish to draw your attention to a particular part of a code block, the relevant lines or items are set in bold: "First, import all the relevant libraries:"

```
>>> from Bio import SeqIO
```

Any command-line input or output is written as follows:

```
>>> from Bio import SeqIO
```

Bold: Indicates a new term, an important word, or words that you see onscreen. For instance, words in menus or dialog boxes appear in **bold**. Here is an example: "In the **Create the default IAM role** pop-up window, select **Any S3 bucket**."

> **Tips or important notes**
> Appear like this.

Get in touch

Feedback from our readers is always welcome.

General feedback: If you have questions about any aspect of this book, email us at customercare@packtpub.com and mention the book title in the subject of your message.

Errata: Although we have taken every care to ensure the accuracy of our content, mistakes do happen. If you have found a mistake in this book, we would be grateful if you would report this to us. Please visit www.packtpub.com/support/errata and fill in the form.

Piracy: If you come across any illegal copies of our works in any form on the internet, we would be grateful if you would provide us with the location address or website name. Please contact us at copyright@packt.com with a link to the material.

If you are interested in becoming an author: If there is a topic that you have expertise in and you are interested in either writing or contributing to a book, please visit authors.packtpub.com.

Reviews

Please leave a review. Once you have read and used this book, why not leave a review on the site that you purchased it from? Potential readers can then see and use your unbiased opinion to make purchase decisions, we at Packt can understand what you think about our products, and our authors can see your feedback on their book. Thank you!

For more information about Packt, please visit packt.com.

Share Your Thoughts

Once you've read *Deep Learning for Genomics*, we'd love to hear your thoughts! Scan the QR code below to go straight to the Amazon review page for this book and share your feedback.

https://packt.link/r/1-804-61544-7

Your review is important to us and the tech community and will help us make sure we're delivering excellent quality content.

Download a free PDF copy of this book

Thanks for purchasing this book!

Do you like to read on the go but are unable to carry your print books everywhere?

Is your eBook purchase not compatible with the device of your choice?

Don't worry, now with every Packt book you get a DRM-free PDF version of that book at no cost.

Read anywhere, any place, on any device. Search, copy, and paste code from your favorite technical books directly into your application.

The perks don't stop there, you can get exclusive access to discounts, newsletters, and great free content in your inbox daily

Follow these simple steps to get the benefits:

1. Scan the QR code or visit the link below

https://packt.link/free-ebook/9781804615447

2. Submit your proof of purchase

3. That's it! We'll send your free PDF and other benefits to your email directly

Part 1 –
Machine Learning in Genomics

This part will describe genomics data analysis and machine learning approaches to genomics. You will use state-of-the-art machine learning methods to transform raw genomics data into insights utilizing real-life examples in the life sciences and biotechnology industries.

This section comprises the following chapters:

- *Chapter 1, Introducing Machine Learning for Genomics*
- *Chapter 2, Genomics Data Analysis*
- *Chapter 3, Machine Learning Methods for Genomic Applications*

1
Introducing Machine Learning for Genomics

Machine learning (ML) is the field of science that deals with developing computer algorithms and models that can perform certain tasks without explicitly programming them. This is to say, it teaches the machines to "learn" rather than specifying "rules" from input data provided to them. The machine then can convert that learning into expertise or knowledge and use that for predictions. ML is an important tool for leveraging technologies around **artificial intelligence** (**AI**), a subfield of computer science that aims to perform tasks automatically that we, as humans, are naturally good at. ML is an important aspect of all modern businesses and research. The adoption of ML for genomics applications is changing recently because of the availability of large genomic datasets, improvement in algorithms, and, most importantly, superior computational power. More and more scientific research organizations and industries are expanding the use of ML across vast volumes of genomic data for predictive diagnostics, as well as to get biological insights at the scale of population health.

Genomics, the study of the genetic constitution of organisms, holds promise in understanding and diagnosing human diseases or improving our agriculture and livestock. The field of genomics has seen exponential growth in the last 15 years, mainly due to recent technological advances in **High-throughput sequencing also known as next-generation sequencing** (**NGS**) technologies generating exponential amounts of genomics data. It is estimated that between 100 million and as many as 2 billion human genomes could be sequenced by 2025 (`https://journals.plos.org/plosbiology/article?id=10.1371/journal.pbio.1002195`), representing an astounding growth of four to five orders of magnitude in 10 years and far exceeding the growth of many big data domains. This complexity and the sheer amount of data generated create roadblocks not only to the acquisition, storage, and distribution but also to genomic data analysis. The current tools used in the genomic analysis are built on top of deterministic approaches and rely on rules encoded to perform a particular task. To keep up with data growth, we need more and new innovative approaches, such as ML, in genomics to enrich our understanding of basic biology and subject them to applied research. In this chapter, we'll learn what ML is, why ML is essential for genomics, and what value ML brings to life sciences and biotechnology industries that leverage genome data for the development of genomic-based products. By the end of this chapter, you will understand the limitations of the current conventional algorithms for genomic data analysis, how solving problems with ML is different from conventional approaches, and how ML approaches can fill in those gaps and make generating biological insights very easy.

As such, in this chapter, we're going to cover the following main topics:

- What is machine learning?
- Why machine learning for genomics?
- Machine learning for genomics in life sciences and biotechnology

What is machine learning?

Before we talk about ML, let's understand what AI is. In the simplest terms, AI is the ability of a machine to mimic human intelligence and iteratively improve itself based on the information it collects. The goal of AI is to build systems to perform actions that are routinely done by humans such as problem-solving, pattern matching, image recognition, knowledge acquisition, and so on. ML, a subset of AI, is the process of training a model to learn and improve from experience. **Deep learning (DL)**, in turn, is a subfield of ML, in which we leverage **artificial neural networks (ANNs)** to mimic the human brain and find the nonlinear relationships between the input and output to generate predictions (*Figure 1.1*):

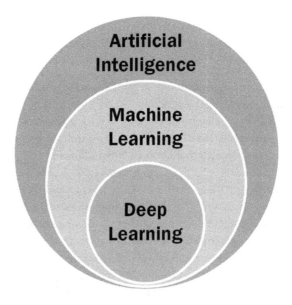

Figure 1.1 – AI versus ML versus DL – how they are related

In ML, a model is built based on input data and an underlying algorithm to make useful predictions from real-world data. In a simplified ML, "features" that represent an individual measurable property of the data are provided as input, and "labels" are returned as the predictions. Suppose we want to predict whether a particular sequence of DNA has a binding site for a **transcription factor** (**TF**) of your interest or not. Using the traditional approach, we would use a **positional weight matrix** (**PWF**) to scan the sequence and identify the potential motifs that are overrepresented. Even though this works, this is extremely difficult, manual, scalable, and so on. Using an ML-based approach, we would give an ML model plenty of DNA sequences until the ML model learns the mathematical relationship between the features from those DNA sequences that either have or don't have binding sites (labels) based on experimental results. It then uses this knowledge to make decisions on new data and make informed predictions. For example, we could give the ML model an unknown DNA sequence, and it would predict the correct binding site motif if present. This is one such example of why ML is a good fit for genomics problems. Some other ways in which ML can be used in genomics include identifying genetic disorders, predicting the type of cancer from genetic variants, improving disease prognosis, and so on.

Why machine learning for genomics?

One of the most important events in the field of biology was the completion of the human genome sequence in 2003, which is considered one of the significant milestones in genomics. Since then, genomics has been evolving rapidly, from research to clinical practice at scale, especially in oncology and infectious diseases. Genomics, because of its ability to identify root causes of diseases due to tiny changes in the genome, fueled the discovery of many important disease genes – particularly rare disease genes – which brought clinical decision-making one step closer to personalized medicine. As a result, sequencing efforts have exploded globally, and so the amount of genomics data that's being generated has shot up. Along with sequencing efforts, biological techniques have started to increase in complexity and number, resulting in large-scale genomics data being generated. It is estimated that there will be between 2 and 40 exabytes of genomics data generated in the next decade (https://www.ncbi.nlm.nih.gov/pmc/articles/PMC4494865/). This is a lot of data, which the current computational and bioinformatics tools can handle, extract, interpret, and identify biological insights. ML, with its inherent nature of learning from experience, holds incredible promise in analyzing this large and complex genomic data. Since ML algorithms can detect patterns in the data automatically, it is suitable for interpreting this large trove of genomic data.

ML has a strong place in genomics since it uses mathematical and data analysis techniques that are applied to complex multi-dimensional datasets, such as genomic datasets, to build predictive models and uncover insights from those models. ML can transform heterogeneous and large-scale genomic datasets into biological insights. ML approaches rely on sophisticated statistical and computational algorithms to make biological predictions. It does this by mapping the complex association between the input features and the labels or finding complex patterns in the input features and creating groups of samples based on similarities using supervised and unsupervised methods, respectively. They can learn useful and new patterns from data that is hard to find by experts. There is now a huge demand for applying ML to genomic datasets because of their huge success in other domains.

Machine learning for genomics in life sciences and biotechnology

Because of the incredible promise that ML has shown for genomics applications such as drug discovery, diagnostics, precision medicine, agriculture, and biological research, more and more life science and biotech organizations are leveraging ML to analyze genomic data for population health and predictive analytics. As per the market research study, which takes into account technology, functionality, application, and region, the global AI in the genomics market is forecasted to reach $1.671 billion by 2025 from $202 million in 2020 (`https://www.marketsandmarkets.com/Market-Reports/artificial-intelligence-in-genomics-market-36649899.html`). The main drivers for this growth can be attributed to the need to control spiraling drug costs, increasing public and private investments, and, most importantly, the adoption of AI solutions in precision medicine. The recent COVID-19 pandemic has played its part in accelerating the adoption of AI for genomics as well (`https://www.jmir.org/2021/3/e22453/`). Even though the outlook for ML for genomics is exciting, there is a lack of a skilled workforce to develop, manage, and apply these ML methodologies in genomics. Additionally, integrating these ML systems into existing systems is a challenging task that requires a proper understanding of the concepts and techniques. For researchers to stand out from the crowd and contribute to data-driven decisions by the company, they must have the necessary skill set.

This book will address the problem of the skill gap that currently exists in the market. This book is a Swiss Army knife for any research professional, data scientist, or manager who is getting started with genomic data analysis using ML. This book highlights the power of ML approaches in handling genomics big data by introducing key concepts, employing real-life business examples, use cases, best practices, and so on to help fill the gaps in both the technical skill set as well as general mentality within the field.

Exploring machine learning software

Before we start the tutorials, we will need some tools. To accommodate users regarding their specific operating system requirements, we will use ML software that is compatible across all operating systems, whether it's Windows, macOS, or Linux. We will be using Python programming language and the Python libraries such as BioPython for genomic data analysis, Scikit-learn for ML building, and Keras to train our DL models. Let's take a closer look at these pieces of ML software.

Python programming language

We will be using the Python programming language throughout this book. Python is a widely used programming language for researchers because of its popularity, the available packages that support all types of data analysis, and its user-friendliness. More importantly, ML, DL, and the genomic community routinely use Python for their own analysis needs. Throughout this book, we will use Python version 3.7 and look at a few ways of installing Python using Pip, Conda, and Anaconda.

Visualization

We will be using the Matplotlib and Seaborn Python packages, which are the two most popular visualization libraries in Python. They are quick to install, easy to use, and easy to import in the Python script. They both come with a variety of functions and methods to use on the data. Throughout this book, we will use *Matplotlib version 3.5.1* and *Seaborn version 0.11.2*. We will look at a few ways of installing these libraries in the subsequent chapters.

Biopython

We will also be using Biopython, a Python module that provides a collection of Python tools for processing genomic data. It creates high-quality, reusable calls for analyzing complex genomic data. It has inherent libraries to connect to databases such as Swiss-Port, NCBI, ENSEMBL, and so on. We will use Biopython version 1.78 and look at separate ways of installing Biopython using Pip, Conda, and Anaconda.

Scikit-learn

Scikit-learn is a Python package written for the sole purpose of performing ML and is one of the most popular ML libraries used by data scientists. It has a rich collection of ML algorithms, extensive tutorials, good documentation, and, most importantly, an excellent user community. For this introductory chapter, we will use scikit-learn for developing ML models in Python. Wherever applicable, we will use scikit-learn version 1.0.2 and look at separate ways of installing scikit-learn in the subsequent chapters.

Summary

In this first chapter, you were introduced to the concept of ML for genomics. We gained a brief understanding of ML in several genomic applications in the life science, pharma, clinical, and biotechnology industries. We also looked at the rapid strides that NGS has made in the last 15 years and how it contributed to the production of genomic big data. Then, we understood how ML can be used to analyze genomic data for the development of genomic-based products.

Finally, we looked at the different programming languages, including the most popular genomic library and ML software that we will be using throughout this book. You will mainly use Python and scikit-learn for developing models, Biopython for genomic data analysis, and some open source tools for model training and productionalizing them for deploying models.

In the next chapter, we will introduce the fundamentals of genomic data analysis.

2
Genomics Data Analysis

Genomics gained mainstream attention when the **Human Genome Project** published the complete sequence of the human genome in 2003. Over the last decade, genomics has become the backbone of drug discovery, targeted therapeutics, disease diagnosis, and precision medicine, leading to the chances of successful clinical trials. For example, in 2021, over 33% of FDA-approved new drug approvals were personalized medicines, a trend that sustained for the past five years (`https://www.foley.com/en/insights/publications/2022/03/personalized-medicine-2021-fda-guideposts-progress`). This growing use of genomics can be mainly attributed to the drastic decrease in the cost and turnaround time of DNA sequencing. For instance, while human genome sequencing was reported to cost around $3 billion and took 13 years to complete, today, you can get your genome sequenced in a day with less than $200 (`https://www.medtechdive.com/news/illumina-ushers-in-200-genome-with-the-launch-of-new-sequencers/633133/`). Because of this incredible success of the genomics industry, more and more research professionals and scientists are now routinely generating genomic data than ever before to understand how genome functions affect human health and disease. It is estimated that over 100 million genomes will be sequenced by 2025 (`https://www.biorxiv.org/content/10.1101/203554v1`) and with the right data analysis and interpretation of this massive data, this information could pave the way to a new golden age of precision medicine.

To find and interpret biological information hidden within this data, it is important to have a solid foundation of genomics data analysis methods and algorithms. The objective of this chapter is to provide fundamentals of genomics data analysis using Biopython, which is one of the most popular Python modules that provides a suite of new commands for working with sequence data. Using Biopython, you can make sense of the vast amounts of available genomics data and extract biological insights. If you are a genomic scientist or a researcher working in the area of genome biology, or someone familiar with these concepts, then please free to skip this chapter or quickly skim through it for a quick refresh. By the end of this chapter, you will know the fundamentals of genomics, genome sequencing, and genome data analysis, how to use the Biopython module of Python for genomics data analysis, and prepare the data in such a way that is compatible with **machine learning** (**ML**).

As such, the following topics are covered in this chapter:

- What is a genome?
- Genome sequencing
- Analysis of genomic data
- Introduction to Biopython for genomic data analysis

Technical requirements

This course assumes that you have a basic knowledge of Python for programming, so we will not introduce Python in this book.

> **Note**
>
> For a quick refresher on Python fundamentals, please refer to `https://www.freecodecamp.org/news/python-fundamentals-for-data-science/`.

Instead, you will be introduced to Biopython, which is a powerful library in Python that has tools for computational molecular biology for performing genomics data analysis.

Installing Biopython

Installing Biopython is very easy, and it will not take more than a few minutes on any operating system.

Step 1 – Verifying Python installation

Before we install Biopython, first check to see whether Python is installed using the following command in your command prompt:

```
$ python --version
Python 3.7.4
```

> **Note**
> The $ character represents the command prompt.

If your command prompt returns something like this, then it shows that Python is installed and *3.7.4* is your version of Python on your computer.

> **Note**
>
> Biopython only works with Python version *2.5* and above. If your Python version is *<2.5*, you should upgrade your Python version.

Alternatively, if you get an error like this, then you should download the latest version of Python from `https://www.python.org/downloads/`, install it, and then run the preceding command again:

```
-bash: python: command not found
```

Step 2 – Installing Biopython using pip

The easiest way to install Biopython is through the `pip` package manager, and the command to install is as follows:

```
$ pip install Biopython==1.79
Collecting Biopython
  Using cached https://files.pythonhosted.org/
packages/4a/28/19014d35446bb00b6783f098eb86f24440b9c099b1f1ded3
3814f48afbea/Biopython-1.79-cp37-cp37m-macosx_10_9_x86_64.whl
Collecting numpy (from Biopython)
  Downloading https://files.pythonhosted.org/packages/09/8c/
ae037b8643aaa405b666c167f48550c1ce6b7c589fe5540de6d83e5931ca/
numpy-1.21.5-cp37-cp37m-macosx_10_9_x86_64.whl (16.9MB)
    100% |████████████████████████████| 16.9MB 3.4MB/s
Installing collected packages: numpy, Biopython
Successfully installed Biopython-1.79 NumPy-1.21.5
```

The preceding response indicates that Biopython version *1.79* is successfully installed on your computer.

If you have an older version of Biopython, try running the following command to update the Biopython version:

```
$ pip install Biopython –upgrade
Requirement already up-to-date: Biopython in python3.7/site-
packages (1.79)
Requirement already satisfied, skipping upgrade: numpy in
python3.7/site-packages (from Biopython) (1.21.5)
```

The preceding message indicates that Biopython is the latest version that you have on your computer. If your Biopython version is older than the most recently updated one of Biopython, then the old version of Biopython and NumPy (a dependency for Biopython) will be replaced by the new version.

Step 3 – Verifying Biopython installation

After you have successfully installed Biopython, you can verify Biopython installation on your machine by running the following command in your Python console:

```
$ python3.7
Python 3.7.4 (default, Aug 13 2019, 15:17:50)
[Clang 4.0.1 (tags/RELEASE_401/final)] :: Anaconda, Inc. on
darwin
Type "help", "copyright", "credits" or "license" for more
information.
>>> import Bio
>>> print(Bio.__version__)
1.79
```

> **Note**
>
> >>> here represents the Python prompt where you would enter code expressions. Please note the underscores before and after the version. You can exit the console by using the exit() command or pressing *Ctrl + D* (which works in Linux and macOS only).

The preceding output shows the version of Biopython, which, in this case, is *1.79*. If that command fails, your Biopython version is *very* out of date and you should upgrade it to the new version as indicated before.

There are alternate ways of installing Biopython, such as installing from the source, and you will find more information on it here: https://biopython.org/wiki/Packages

Matplotlib

We will be using Matplotlib, a very popular Python library for visualization. It is one of the easiest libraries to install and use. To install Matplotlib, simply run pip install matplotlib in your terminal. Then, you can include import matplotlib.pyplot as plt in your Python script, which you will see later in the hands-on section of this chapter.

What is a genome?

Before we discuss genomes, let's do a quick Genetics 101. A **cell** represents the fundamental structural and functional unit of life. **DNA** contains the instructions that are needed to perform different activities of the cell. DNA is the basis of genetic studies and consists of four building blocks called nucleotides – **adenine (A)**, **guanine (G)**, **cytosine (C)**, and **thymine (T)**, which store information about life. The sequence of DNA is a string of these building blocks, also referred to as **bases**. DNA has a double-helix structure with two complementary polymers interlaced with each other. In the complementary strand of DNA, A matches with T, and G matches with C, to form base pairs.

A genome represents the full DNA sequence of a cell that contains all the hereditary information. The genome consists of information that is needed to build and maintain the whole living organism. The size of genomes is different from species to species. For example, the human genome is made up of 3 billion base pairs spread across 46 chromosomes, whereas the bread wheat genome consists of 42 chromosomes and ~ 17 gigabases. A region of a genome that transcribes into a functional RNA molecule, or transcribes into an RNA and then encodes a functional protein, is called a **gene**. This is the simplest of definitions, and there are several definitions of a gene, but overall, a gene constitutes the fundamental unit of heredity of a living organism. By analogy, you can imagine the four nucleotides (A, G, C, and T) that make up the gene as letters in a sentence, genes as sentences in a book, and the genome as the actual book consisting of tens of thousands of words.

Genome sequencing

After the discovery of the DNA structure, scientists were curious to determine the exact sequence of DNA (aka interpreting the whole book). A lot of pioneering discoveries paved the way for the sequencing of DNA, starting with Walter Gilbert when he published the first nucleotide sequence of the DNA *lac* operator consisting of 24 base pairs in 1973 (`https://www.ncbi.nlm.nih.gov/pmc/articles/PMC427284`). This was followed by Frederick Sanger who for the first time sequenced the complete DNA genome of the phi X174 bacteriophage (`https://pubmed.ncbi.nlm.nih.gov/731693/`). Sanger pioneered the first-ever sequencing of genes through the method of DNA sequencing with chain-terminating inhibitors. In the last 50 years, available sequencing technologies have been restricted to relatively small genomes, but advances in DNA sequencing technologies such as **next-generation sequencing** (**NGS**) have revolutionized genome sequencing because of their cost, speed, throughput, and accuracy. As more and more genomes are sequenced, this extended knowledge can be utilized for the development of personalized medicines to prevent, diagnose, and treat diseases and, ultimately, support clinical decision-making and healthcare. It's beyond the scope of this chapter to go into the historical background of DNA sequencing technologies. Here, we will briefly present the different DNA sequencing technologies that have led to major milestones in genome sequencing.

Sanger sequencing of nucleic acids

Sanger sequencing, also known as the "chain termination method" is the first-generation sequencing method, and was developed by Frederick Sanger and his colleagues in 1977 (`https://www.ncbi.nlm.nih.gov/pmc/articles/PMC431765/`). It was used by the **Human Genome Project** to sequence the first draft of the human genome. This method relies on the natural method of DNA replication. Sanger sequencing involves the random incorporation of bases called **deoxyribonucleoside triphosphates** (**dNTPs**) by DNA polymerase into strands during copying, resulting in the termination of *invitro* transcription. These bases in the short fragments are then read out based on the presence of a dye molecule attached to the special bases at the end of each fragment. Despite being replaced by NGS technologies, Sanger sequencing remains the gold standard sequencing method and is routinely being used by research labs throughout the world for quick verification of short sequences generated by **polymerase chain reaction** (**PCR**) and other methods.

Evolution of next-generation sequencing

Next-generation sequencing first became available at the start of the 21st century and it completely transformed biological sciences, allowing any research lab to perform sequencing for a wide variety of applications and study biological systems at a level unimaginable before. NGS aims to determine the order of nucleotides in the entire genome or targeted regions of DNA or RNA in a high-throughput and massively parallel manner. The biggest advance that NGS offered compared to traditional Sanger sequencing and other approaches is the ability to produce ultra-high-throughput genomic data at a scale and speed. NGS technologies can be broadly divided into second-generation sequencing and third-generation sequencing technologies, depending on the type of sequencing methodology.

Second-generation DNA sequencing technologies

Alongside the developments in large-scale dideoxy-sequencing efforts, another sequencing technology that emerged in the last 15 years that revolutionized DNA sequencing is the NGS technology. There are a lot of technologies under NGS starting with pyrosequencing, by 454 Life Sciences, but the most important of all of them are bridge amplification technologies that were brought forth by Solexa, which was later acquired by Illumina. In contrast to Sanger sequencing, Illumina leverages **sequencing by synthesis** (**SBS**) technology, which is the tracking of labeled nucleotides as the DNA is copied in a massively parallel fashion generating output ranging from 300 kilobases up to several terabases in a single run. The typical size fragments generated by Illumina are in the range of 50-300 bases. Illumina dominates the NGS market.

Third-generation DNA sequencing technologies

These technologies are capable of sequencing single DNA molecules full length without amplification, and they allow the production of sequences (also called as reads) much longer than second-generation sequencing technologies such as Illumina. Pacific Bioscience and Oxford Nanopore Technologies dominate this sector with their systems called **single-molecule sequencing real-time** (**SMRT**) sequencing and **nanopore sequencing**, respectively. Each of these technologies can rapidly generate very long reads of up to 15,000 bases long from single molecules of DNA and RNA.

Analysis of genomic data

Genomic data analysis aims to provide biological interpretation and insights from genomics data and help drive innovation. It is like any other data analysis with the exception that it requires domain-specific knowledge and tools. With the advances in NGS technologies, it is estimated that genomic research will generate significant amounts of data in the next decade. However, our ability to mine insights from this big data is lagging behind the pace at which the current data is being generated. As new and more high-throughput genomic data is getting generated, data analysis capabilities are sought-after features for researchers and other scientific professionals in both academia and industry.

Steps in genomics data analysis

Genomics data is generally complex in nature and size. Researchers are currently facing an exciting yet challenging time with this available data that needs to be analyzed and understood. Analyzing this big genomics data can be extremely challenging. A natural course of action can be ML, which is fast becoming the go-to method for analyzing this data to mine biological insights. We will discuss the application of ML for genomics data in the next chapter but for now, let's understand the different steps involved in the analysis of genomics data. The main aim of genomics data analysis is to do a biological interpretation of large volumes of raw genomics data. It is very similar to any other kind of data analysis but in this case, it often requires domain-specific knowledge and tools. Here, we will discuss data analysis steps for analyzing genomics data. Typical analysis of genomic data consists of multiple interdependent **extract, transform, and load** (ETL) that transform raw genomic data from sequencing machines into more meaningful downstream data used by researchers and clinicians.

A typical genomics data analysis workflow consists of the following steps: data collection, data cleaning, data processing, **exploratory data analysis** (EDA), modeling, visualization, and reporting. You might have seen several versions of this workflow but ultimately, a typical genomic data analysis workflow boils down to these five main steps – raw data collection, transforming the data, **Exploratory Data Analysis (EDA)**, modeling (statistical or machine learning), and biological interpretation, as shown in *Figure 2.1*:

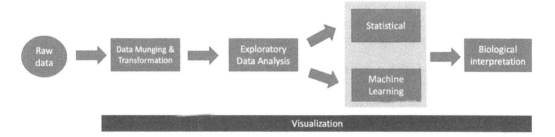

Figure 2.1 – Steps in a typical genomic data analysis workflow

Even though the steps in the genomic data analysis are linear, there are instances when you go back and repeat many of these steps to answer some questions related to data quality or add new datasets in the analysis or optimize the parameters.

Data collection

Data collection refers to any source, experiment, or survey that provides raw data. The genomic data analysis starts right after the experimentation stage, where data is generated if no data is available or if the question cannot be solved with available data. Many times, you don't need to generate the data and instead can use publicly available datasets and specialized databases. The type and amount of data that needs to be collected are entirely dependent on the questions along with the technical and biological variability of the experimental system under study.

Data transformation

Data transformation consists of converting raw data into data that can be used for downstream analysis, such as EDA and modeling. It consists of quality control and data processing.

Quality control and cleaning

The quality control and cleaning step aims to identify the quality issues in the data and clean them from the raw dataset, generating high-quality data for modeling purposes. Quite frequently, the analysis starts with the raw genomics data (processed data if you are lucky), and it's mostly messy data. Like any other raw data, genomic data consists of missing values, outliers, invalid, and other noisy data. Since it is a well-known fact that the quality of the output is determined by the quality of the input (garbage in, garbage out), it is super important to subject the data to quality control and cleaning before using it for downstream analysis.

Data processing

The goal of data processing is to convert raw data into a format that is amenable to visualization or EDA or modeling. The data processing step starts after the raw data was collected and cleaned. Briefly, the data processing steps include the following:

- Data munging, which is transforming data from one format to another format

- Data transformation, which includes either normalizing the data or log transformation of the data

- Data filtering, which is a key step to remove data points that have unreliable measurements or missing values.

Exploratory data analysis

EDA is the most crucial step of any data analysis. The goal of EDA is to use various statistics and visualization methods on either the processed or semi-processed data to find the relationships between the variables in the input data, or between samples or correlations between input and output, or detect outliers from the data, and so on. Some examples of EDA include nucleotide composition, di-nucleotide frequencies, **guanine-cytosine (GC)** content, amino-acid frequencies, and so on.

Modeling

Post-EDA, you might be interested in building a model to get the outcome or get some biological insights from the processed data. Modeling is the process of mapping the relationship between input variables and output. For example, modeling to predict the disease status of a patient either diseased or not from the genomic data such as gene expression, variants detected, and so on. This is called **predictive modeling** and we will discuss the different algorithms for performing predictive modeling in genomics in the next chapter. As you can see from *Figure 2.1*, there is also statistical modeling where we can use statistical methods. An example of statistical modeling in genomics is differential gene expression analysis where we are interested to find whether a gene or set of genes is significantly different between datasets, for example, non-diseased versus diseased patients.

Visualization and reporting

Visualization is not a standalone step but is necessary for all the preceding steps. It is visualizing the data using visualization tools before and after each step of the workflow to make informed decisions. Reporting is compiling all the preceding information through plots, tables, and descriptions that describe the complete data analysis. In genomics, there are innumerable visualization and reporting methods developed, and we will use some of them for genomic data analysis later in the hands-on section of this chapter.

Cloud computing for genomics data analysis

The incredible success of NGS technologies with their ability to generate large volumes of data at a low cost and speed has created several challenges in terms of the acquisition, storage, distribution, and data analysis. For example, just a single human genome sequencing run produces approximately 200 GB of raw data and an additional 100 GB of data per genome after data analysis. If we were to manage to sequence 100 million genomes by 2025, then we will accumulate over 200 Exabyte of raw data. In addition to genomics data storage, the data analysis and interpretation of this massive amount of data are incredibly challenging because of the high-throughput computing needed to process this data. This is where cloud computing technologies can aid genomics. Cloud computing has completely transformed the way we run computing in various domains of life sciences. Cloud computing can transform the existing small or on-premises genomic data analysis into efficient genomics data analysis systems on the cloud. They offer solutions for research professionals who currently lack the tools to make full use of this data. The demand for cloud-based solutions for genomics was evident in the last several years, with several big players coming and supporting genomics, such as **Amazon Web Services** (**AWS**), Azure, and **Google Cloud Platform** (**GCP**). Even though cloud computing provides the resources for computation, they are still expensive and require a credit card to register to use its services. So, to address that, we will either run the analysis on laptops or use free open-source computing platforms such as Google colabs (which you will learn about in future chapters). For this chapter though we will run the analysis on our computers.

Introduction to Biopython for genomic data analysis

In this section, you will be familiarized with the basics of Biopython, and in the subsequent section, you will use Biopython for solving a real-world research question in genomics.

What is Biopython?

Biopython is a popular Python package developed by Chapman and Chang, mainly intended for biological researchers and data miners to analyze genomic data. It was written mainly in Python but also has support for C code to optimize complex computations. It can be run on any operating system (Windows, Linux, and macOS). Biopython provides lots of functionalities to support genomic data and it makes it easy to use Python for genomic data analysis through reusable modules and classes. In addition to providing basic and advanced genomic functionalities, it also has support for parsers for various popular bioinformatics file formats such as BLAST, ClustalW, FASTA, and GenBank, as well as support online databases and servers such as NCBI, Expasy, and so on.

Quick start of Biopython

Before we dive deep into genomics use cases, let's have a quick introduction to Biopython. If you are already familiar with Biopython basics, please feel free to skip this section and jump into the hands-on section.

Working with sequences

As genomics is the field of sequences, you cannot get away without using them. A sequence is a series of letters, for example, A, G, C, and T string together in DNA, or A, G, C, and U string together in RNA, or amino acids D, T, S, E, P, G, A, C, V, M, I, L, Y, F, H, K, R, W, Q, and N string together in a protein. Let's look at how Biopython deals with sequence data but before that, let's refresh your Python knowledge briefly.

> **Note**
> An object is defined as a collection of data (variables) and the methods (functions) that define and act on that data. In Python, everything is an object.

Seq object

The first and the most important object to deal with sequence data in Biopython is the Seq object. It essentially combines a Python string with biological methods such as DNA, RNA, or protein.

For example, you can create a Seq object in Biopython for a DNA sequence consisting of letters such as 'AGTAGGACAGAT', as follows:

```
>>> from Bio.Seq import Seq
>>> my_seq = Seq("AGTAGGACAGAT")
```

```
>>> my_seq
Seq('AGTAGGACAGAT')
>>> print(my_seq)
AGTAGGACAGAT
```

You may be wondering how the `Seq` object that you created in the preceding code is different from the regular Python string object since they look the same. Well, it differs from the Python string in the kinds of methods it supports. For example, the regular Python string method does not support `complement` and `reverse_complement` functions, whereas the Biopython `Seq` object supports those two along with several other very useful methods. Some of the popular methods for `Seq` objects include `complement` and `reverse_complement`, as shown in the following code block:

```
>>> my_seq.complement()
Seq('TCATCCTGTCTA')
>>> my_seq.reverse_complement()
Seq('ATCTGTCCTACT')
```

As you can see from the preceding code block, the Biopython `Seq` object is very convenient to apply these methods without you writing a function.

Each `Seq` object has two important attributes:

- Data – The actual sequence string (`'ATCTGTCCTACT'`).

- Alphabet – The type of sequence, for example, DNA, RNA, or protein. By default, it doesn't represent any sequence. That means Biopython doesn't necessarily know whether the input to the `Seq` object is a nucleotide (A, G, C, or T) or protein sequence consisting of the amino acids alanine (A), cysteine (C), glycine (G), and threonine (T). So, keep this in mind when you call methods such as `complement` and `reverse_complement`, depending on the type of input sequence.

The `Seq` object supports two types of methods – general methods (`find`, `count`, and so on) and nucleotide methods (`complement`, `reverse_complement`, `transcribe`, `back_transcribe`, `translate`, and so on). We will use a few of these methods later in the chapter.

SeqRecord object

After the `Seq` object, the next most important object in Biopython is `SeqRecord` or **sequence record**. This object differs from the `Seq` object in that it holds a sequence (as a `Seq` object) with additional information such as identifier, name, and description. You can use the `Bio.SeqIO` module with the `SeqRecord` object in Biopython.

SeqIO object

The SeqIO object in Biopython provides the standard sequence input/output interface. It supports several file formats as input and output. The following table (*Figure 2.2*) lists the selected file formats that the Bio.SeqIO module can read, write, and index. For a full list of file formats, please refer to Biopython's documentation (https://biopython.org/wiki/Documentation).

Format name	Notes
FASTA	This format refers to the input data where each record starts with a ">" line in the first line followed by the actual sequence in multiple lines.
FASTA-2line	Similar to the FASTA format but has no line wrapping and exactly two lines per record.
FASTQ	This is the popular format for data coming out of sequencing instruments such as Illumina. Just like FASTA files, it includes the header and a sequence but in addition, it also includes sequencing qualities.
GenBank or GB	This is the format from the NCBI database. It is a list of shaped records containing detailed information about DNA samples such as locus, organism, type of sequence, source of sequence, and so on.

Figure 2.2 – List of popular file formats that are supported in Biopython

The main function in the SeqIO object is parse, which accepts either a file handle or a filename along with a format name. It then returns a SeqRecord iterator. Let's see the parse function in action using FASTA and GenBank file formats now.

FASTA

If you are a genomic researcher, then there is a very high chance that you have used the FASTA format before or at least heard about it. Let's use the Bio.SeqIO module's parse function on an example FASTA file.

But first, let's import the SeqIO model into the Python console:

```
>>> from Bio import SeqIO
```

The `parse()` method takes two arguments – the file handle and the file format. In this example, we will provide the file handle as `fh_in` and the file format as `fasta`:

```
>>> with open("example.fasta") as fh_in:
...       for record in SeqIO.parse(fh_in, "fasta"):
...               print(record.id)
```

This returns an iterable object, `record`, with `SeqRecord` on every iteration, which provides a lot of sophisticated and easy methods. We will see this in action in the sample use case section coming up.

GenBank

Unlike the FASTA format, GenBank is a sequence format that has rich information about gene sequences and includes several fields for various kinds of annotation. The only difference between this file format compared to the FASTA file format while loading into the `parse` function is to change the format option in the `parse` method:

```
>>> with open("example.gbk") as fh_in:
...       for record in SeqIO.parse(fh_in, "genbank"):
...               print(record.annotations)
```

As shown, we changed the formation from `"fasta"` to `"GenBank"` while parsing a GenBank format file.

We will see an example of this in the sample use cases section.

Genomic data analysis use case – Sequence analysis of Covid-19

Now that you have gained some basic knowledge of Biopython and how it can be used on sequence data of various file formats, let's use this knowledge in a real-world research problem specifically related to sequence analysis. The goal here is to introduce Biopython with the genomic context in the background. The sample use case listed here is to analyze genomic data of **Covid-19** with Biopython. We believe tailoring material to the context of genomics makes a difference when learning Biopython for the sake of analyzing genomic data.

Sequence analysis is considered an important part of genomic data analysis. An example of sequence analysis includes inferring sequence composition, calculating GC content, calculation of % *T*, % *A*, and so on. In addition, more complicated tasks such as motif searching are also part of sequence analysis. These are considered features derived from sequences for training models, which we will see in the next chapter. These features have a direct impact on prediction purposes during model training. So, it is important to understand how to extract these features from sequence data, and for that, we will use Biopython. For this example, we will use **SARS-CoV-2**, which is a causative agent for Covid-19 and needs no introduction because of the widespread destruction it caused across the whole world causing millions of deaths:

1. The full sequence of the Wuhan virus can be downloaded (`https://www.ncbi.nlm.nih.gov/nuccore/NC_045512.2?report=fasta`); save it as `covid19.fasta` locally on your computer. You can open this FASTA file in a text editor of your choice, and this is how it looks:

>NC_045512.2 Severe acute respiratory syndrome coronavirus 2 isolate Wuhan-Hu-1, complete genome

ATTAAAGGTTTATACCTTCCCAGGTAACAAACCAACCAACTTTCGATCTCTTGTAGATCTGTT
CTCTAAACGAACTTTAAAATCTGTGTGGCTGTCACTCGGCTGCATGCTTAGTGCACTCACGCA
GTATAATTAATAACTAATTACTGTCGTTGACAGGACACGAGTAACTCGTCTATCTTCTGCAGG
CTGCTTACGGTTTCGTCCGTGTTGCAGCCGATCATCAGCACATCTAGGTTTCGTCCGGGTGTG
ACCGAAAGGTAAGATGGAGAGCCTTGTCCCTGGTTTCAACGAGAAAACACACGTCCAACTCAG
TTTGCCTGTTTTACAGGTTCGCGACGTGCTCGTACGTGGCTTTGGAGACTCCGTGGAGGAGGT
CTTATCAGAGGCACGTCAACATCTTAAAGATGGCACTTGTGGCTTAGTAGAAGTTGAAAAAGG
CGTTTTGCCTCAACTTGAACAGCCCTATGTGTTCA...

Figure 2.3 – Wuhan genome sequence (only a fraction of the sequence is shown here)

Looking at the sequence, you might be wondering why I can't just use Python for parsing this FASTA file and do sequence data analysis such as calculating GC content, calculation of % T, % A , and so on. Well, you can certainly do that, but as discussed previously, Biopython already has objects and methods suited for this kind of sequence data analysis, and so we will take advantage of that.

2. Let's create a simple Python script (`covid19_meta.py`) to parse this FASTA file and run some basic sequence analysis:

```python
from Bio import SeqIO
with open("covid19.fasta") as fh_in:
    for record in SeqIO.parse(fh_in, "fasta"):
            print(f'sequence information: {record}')
            print(f'sequence length: {len(record)}')
```

3. Now, you can execute this script by running it as follows at the command prompt in your terminal:

```
$ python3.7 covid19_meta.py
```
The preceding command will generate the following output in the terminal:
```
sequence information: ID: NC_045512.2
```

```
Name: NC_045512.2
Description: NC_045512.2 Severe acute respiratory
syndrome coronavirus 2 isolate Wuhan-Hu-1, complete
genome
Number of features: 0
Seq('ATTAAAGGTTTATACCTTCCCAGGTAACAAACCAACCAACTTTCGATCTCTT
GT...AAA')
sequence length: 29903
```

> **Note:**
> Make sure you run this script in the same directory as the `covid19.fasta` file you saved in *Step 1*.

Now, let's try the same code in the Python console:

1. First import all the relevant libraries:

   ```
   >>> from Bio import SeqIO
   ```

2. Then, run the following code block:

   ```
   >>> with open("covid19.fasta") as fh_in:
   ...        for record in SeqIO.parse(fh_in, "fasta"):
   ...                print(f'sequence information: {record}')
   ...                print(f'sequence length: {len(record)}')
   ...
   sequence information: ID: NC_045512.2
   Name: NC_045512.2
   Description: NC_045512.2 Severe acute respiratory
   syndrome coronavirus 2 isolate Wuhan-Hu-1, complete
   genome
   Number of features: 0
   Seq('ATTAAAGGTTTATACCTTCCCAGGTAACAAACCAACCAACTTTCGATCTCTT
   GT...AAA')
   sequence length: 29903
   ```

> **Note**
> When you run these commands in the console, make sure you do not include >>>.

So, what's happening here?

1. First, we have loaded the `SeqIO` module from Biopython.

2. Next, we opened the `covid19.fasta` file, as we learned about earlier with the `parse` method.

3. Finally, we printed some basic information about the sequence such as the name, description, number of features, sequence, and sequence length.

What is the length of SARS-COV-2 in kilobases then? Alright, now you know how to read a FASTA file using Biopython. Let's continue using this FASTA file to perform some basic genomic data analysis.

Calculating GC content

GC content is one of the important features of a DNA sequence as it is an important predictor of gene function and species ecology. GC content is calculated by counting the number of Gs and Cs in the sequence and dividing that by the total sequence length.

However, we don't need to write a function to calculate GC; thanks to Biopython, we can load the `Bio.SeqUtils` module and then import the GC function as shown in the following code.

Write a script to calculate the nucleotide percentages as follows:

```
# Covid19_meta1.py
from Bio import SeqIO
from Bio.SeqUtils import GC
with open("covid19.fasta") as fh_in:
    for record in SeqIO.parse(fh_in, "fasta"):
        print(f'GC content: {GC(record.seq)}')}')
$ python3.7 covid19_meta1.py
GC content: 37.97277865097148
```

How can you round the value of GC content to two decimals in Python? The result here shows that the `covid19` sequence has a GC content of ~38. If you are a genomic researcher, then the GC content of `covid19` might look very low to you and in fact, you are right. SARS-CoV-2 has an extremely low GC content compared to other coronavirus species. *Why does this virus have low GC content compared to other viruses in the family? Does this low GC specifically help its survival?* These are some of the interesting questions to explore. Let's move on to calculate the nucleotide content using Biopython now.

Calculating nucleotide content

In addition to GC content, nucleotide content such as the percentages of A, T, C, and G are useful for sequence characterization purposes. Unlike GC content, Biopython doesn't have a function and so let's write our little code to make those calculations using the `count` method.

Write a script to calculate the nucleotide percentages as follows:

```
# Covid19_meta2.py
from Bio import SeqIO
with open("covid19.fasta") as fh_in:
    for record in SeqIO.parse(fh_in, "fasta"):
            seq_record = record.seq
            seq_length = len(record.seq)
            print(f'% of Ts: {round(seq_record.count("T")/seq_
length*100, 2)}')
            print(f'% of As: {round(seq_record.count("A")/seq_
length*100, 2)}')
            print(f'% of Cs: {round(seq_record.count("C")/seq_
length*100, 2)}')
            print(f'% of Gs: {round(seq_record.count("G")/seq_
length*100, 2)}')
```

Executing the script at the command line shows the following output:

```
$ python3.7 covid19_meta2.py
% of Ts: 32.08
% of As: 29.94
% of Cs: 18.37
% of Gs: 19.61
```

Does the percentage of As, Ts, CS, and Gs say anything about the nature of this virus? Unlike GC content, it is not evident what this nucleotide content tells us about this virus.

Dinucleotide content

Even though GC content is important to look at, it's also valuable to count the dinucleotides (AT, AC, GT, and so on) of Covid-19 virus and then compare that with other viruses and see how the Covid-19 sequence is different from other viral sequences:

1. A DNA sequence consists of four different nucleotides (A, G, C, and T). So, first, generate the 16 possible dinucleotides from these four nucleotides and then count the number of each of those dinucleotide sequences from the Covid-19 sequence:

    ```
    # covid19_dinucleotide.py
    from Bio import SeqIO
    import matplotlib.pyplot as plt
    ```

```
nucl = ['A', 'T', 'C', 'G']
di_nucl_dict = {}
with open("covid19.fasta") as fh_in:
    for record in SeqIO.parse(fh_in, "fasta"):
        for n1 in nucl:
            for n2 in nucl:
                di = str(n1) + str(n2)
                di_nucl_dict[di] = record.seq.count(di)
print(di_nucl_dict)
```

This will print the following output when you run:

```
$ python3.7 covid19_dinucleotide.py
{'AA': 2169, 'AT': 2308, 'AC': 2023, 'AG': 1742, 'TA':
2377, 'TT': 2454, 'TC': 1413, 'TG': 2589, 'CA': 2084,
'CT': 2081, 'CC': 784, 'CG': 439, 'GA': 1612, 'GT': 1990,
'GC': 1168, 'GG': 973}
```

The preceding output shows a dictionary consisting of all possible dinucleotides as keys and counts for those dinucleotides as values.

2. EDA and visualization – You can append the following code to the previous script and then can generate the dinucleotide plot:

```
di = [k for k, v in di_nucl_dict.items()]
counts = [v for k, v in di_nucl_dict.items()]
print(di_nucl_dict)
plt.bar(di,counts)
plt.ylabel("Counts")
plt.show()
```

3. Next, for ease of visualization, let's quickly visualize dinucleotide plot and look at their relative distribution concerning each other:

```
$ python3.7 covid19_dinucleotide.py
```

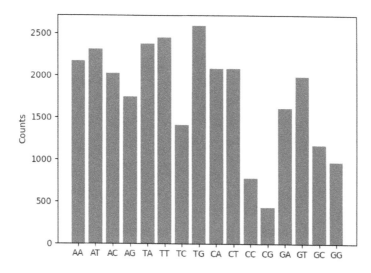

Figure 2.4 – Bar plot showing the dinucleotide count from the Covid-19 sequence

As you can see and expect, the expected dinucleotide counts vary, and now you can repeat the same on different viruses and compare all of them to understand their difference concerning dinucleotide content.

Modeling

So far, you have looked at some of the key steps of sequence analysis, such as data collection, data transformation, EDA, visualization, and so on. The next step would be modeling either using ML-based or statistical-based methods. Since we have the next chapter dedicated to ML-based modeling, we will not go into details of how that is done here but to wrap this real-world use case, we will save all the features that we have extracted so far into a file and get it ready for modeling.

Let's create a Python script and combine all the code that we have written so far, which we can use for generating the features, and use that for modeling in subsequent chapters:

1. First, include the code related to the different features such as dinucleotide content, A, G, C, and T percentage, GC content, and length of the sequence:

```
# covid19_features.py
from Bio import SeqIO
from Bio.SeqUtils import GC
import pandas as pd
nucl = ['A', 'T', 'C', 'G']
final_dict = {}
with open('covid19.fasta') as fh_in:
    with open("test.csv", 'w') as fh_out:
```

```
        for record in SeqIO.parse(fh_in, "fasta"):
            for n1 in nucl:
                for n2 in nucl:
                    di = str(n1) + str(n2)
                    final_dict[di] = record.seq.count(di)
            A_count = record.seq.count('A')
            final_dict['A_count'] = round(A_count/
len(record) * 100, 2)
            C_count = record.seq.count('C')
            final_dict['C_count'] = round(C_count/
len(record) * 100, 2)
            G_count = record.seq.count('G')
            final_dict['G_count'] = round(G_count/
len(record) * 100, 2)
            T_count = record.seq.count('T')
            final_dict['T_count'] = round(T_count/
len(record) * 100, 2)
            final_dict['GC_content'] = round(GC(record.
seq), 2)
            final_dict['Size'] = len(record)
```

All of this should be familiar from the preceding code.

2. Once you have the code for extracting the features, then you are ready to save them to a file using the following code:

```
final_df = pd.DataFrame.from_dict([final_dict])
final_df['virus'] = "Covid19"
final_df.to_csv("covid19_features.csv", index=None)
```

3. Executing the script at the command prompt generates covid19_features.csv:

```
$ python3.7 covid19_features.py
```

Before we end this section, let's print the contents of the output file (covid19_features. csv):

```
$ cat covid19_features.csv
AA,AT,AC,AG,TA,TT,TC,TG,CA,CT,CC,CG,GA,GT,GC,GG,A_
count,C_count,G_count,T_count,GC_content,Size,virus
2169,2308,2023,1742,2377,2454,1413,2589,2084,2081,784,439
,1612,1990,1168,973,29.94,18.37,19.61,32.08,37.97,29903,C
ovid19
```

As you can see, we have combined all the features (A, G, T, and C content, GC content, and dinucleotide content) in a file and added a separate `virus` column to indicate where the features are coming from. Now, you can do the same with a few other viruses and then build a model, evaluate it, and then use that model to predict the type of virus from an unknown sequence. Isn't this powerful?

Motif finder

A **motif** is a pattern in a nucleotide or amino acid sequence that has a specific structure. Sequence motifs play a key role in gene expression regulating both transcriptional and post-transcriptional levels. Motif mining is the foundation of gene function research. Currently, there have been innumerable algorithms including ML- and DL-based motif miners. The goal here is to utilize Biopython's motifs module to access the functionalities of sequence motifs. Let's understand how it is done now:

1. The motif method can be imported from the Biopython library shown as follows:

```
>>> from Bio import motifs >>> from Bio.Seq import Seq
```

2. Let's now create a simple DNA motif object as shown in the following code:

```
>>> my_motif = [Seq("ACGT"), Seq("TCGA"), Seq("CGGC")]
```

Here, we are creating a list of `Seq` objects each containing a four-nucleotide sequence, and passing it to a `my_motif` variable.

3. Now, we will create a motif object from the `seq` object list and print the list:

```
>>> seq = motifs.create(my_motif)
>>> print(seq)
ACGT
TCGA
CGGC
```

As expected, the motif object printed the three different `seq` objects.

4. One of the important attributes that are available for motif objects is `counts`, which prints the counts of each nucleotide at each position. This can also refer as the **position frequency matrix (PFM)**. Printing this matrix is shown as follows:

```
>>> print(seq.counts)
        0       1       2       3
A:    1.00    0.00    0.00    1.00
C:    1.00    2.00    0.00    1.00
G:    0.00    1.00    3.00    0.00
T:    1.00    0.00    0.00    1.00
```

What we see here is a matrix of nucleotide types and their position in the sequence along rows and columns, respectively, and counts in each of the cells.

5.   You can also create a logo from the motif by running the following command:

```
>>> seq.weblogo('my_motif.png')
```

This will generate a `'my_motif.png'` file in the current working directory. Please note that you need to have internet access for this to work.

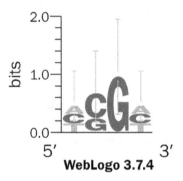

Figure 2.5 – Example logo generated from sequence motif

As you can see, Biopython is powerful in generating this kind of result without you needing to manually write functions to do it. Biopython would be extremely useful in the data processing stage of genomic data analysis, where the goal is to transform the raw data into processed data.

Summary

In this chapter, you were introduced to genomic data analysis for DNA sequencing data. First, we looked at the foundations of genomics and a brief history of DNA sequencing. Next, we explored in depth the genomic data analysis, the different types of data analysis, and a brief introduction to cloud computing for genomics data analysis. Then, we had a quick tutorial on using Biopython for analyzing genomic data. Finally, we applied the Biopython fundamentals to a real-world use case to understand how to perform sequence analysis.

Throughout this book, Python and Biopython are the main languages we will use to analyze genomic data. Biopython is a very powerful Python package for performing genomic data analysis with existing functions and methods that can be used out of the box. Many genomic and Data scientists routinely use Biopython for performing day-to-day jobs writing scripts for parsing the data or creating features for building ML and DL models.

Now that you have gained some understanding of genomic data analysis, let's go ahead and investigate ML for genomics applications in the next chapter.

Machine Learning Methods for Genomic Applications

Have you ever wondered how YouTube recommends videos to you, banks detect fraudulent activity and send notifications to you, or Gmail filters the spam messages from your inbox? These are just a few examples of how the world of business is currently using **machine learning** (ML). The field of ML has impacted numerous areas of modern society and is responsible for some of the most significant improvements in technologies such as self-driving cars, exploring the galaxy, predictions for disease outbreaks, and so on. The enormous growth in ML is primarily driven by its huge success in solving real-world business problems in healthcare, finance, e-commerce, agriculture, life sciences, pharmaceuticals, and biotechnology. The life sciences and biotechnology industries are huge and diverse with many subsectors. Very popular fields are drug discovery and manufacturing, therapeutics, diagnostics, genomics, and so on.

The field of genomics has seen massive growth in the past few years because of advancements in sequencing technologies that have pushed genomic data in the wave of big data. ML techniques, with their ability to analyze large-scale data, can turn this complex genomic data into biological insights that can in turn convert into useful products. ML algorithms have been widely used in biology and genomics to apply to complex multi-dimensional datasets for the building of predictive models to solve complex biological problems such as disorder risk prediction, mental illness prediction, diagnosis, treatment, and so on. This chapter introduces the main ML algorithms and libraries that are commonly used in genomic data analysis. By the end of the chapter, you will know what supervised and unsupervised ML methods are, understand the most common ML algorithms and libraries for genomics, and know when and how to use them. You will know how to build a predictive model using genomics data and be aware of the challenges encountered by ML algorithms for genomics and potential solutions for addressing the same.

As such, here are the contents of this chapter:

- Genomics big data
- Supervised and unsupervised ML

- ML for genomics

- An ML use case for genomics—Disease prediction

- ML challenges in genomics

Technical requirements

Let's understand the technical requirements for the different Python packages and other ML libraries that are needed to apply ML in genomics in this chapter.

Python packages

The following are some common Python packages that every data scientist and genomic researcher uses for not only genomic analysis for any kind of data analysis.

Pandas

`Pandas` is one of the most popular data analysis tools in Python. `Pandas` do not need an introduction as it is part and parcel of every data scientist's tool. The great thing about `Pandas` is it contains all the functions and methods to support data analysis irrespective of the type of data. It's also super easy to install Pandas, which you can do by simply entering `pip install pandas` in your terminal. Then, you can include `import pandas as pd` in your Python script, which you will see later in the chapter.

Matplotlib

We will be using Matplotlib, a very popular Python library for visualization. It is one of the easiest libraries to install and use. To install Matplotlib, simply run `pip install matplotlib` in your terminal. Then, you can include `import matplotlib.pyplot as plt` in your Python script, which you will see later in the hands-on section of this chapter.

Seaborn

Seaborn is another visualization Python library that we will use in this chapter because of its easy-to-use functionalities for plotting. Again, as with any other Python package, it is easy to install (`pip install seaborn`) and easy to include in Python scripting (`import seaborn as sns`).

ML libraries

An ML library is a compilation of functions and routines that are readily available to use without building from scratch. These are an essential part of ML for any field, and they help save time and effort in writing complex code. Furthermore, they also solve complex problems such as data manipulation, text processing, scientific computation, and so on. They simplify the task of performing ML because of their ability to support a variety of ML algorithms. Some of the ML libraries, based on their popularity among ML practitioners and enthusiasts all around the world, include scikit-learn, Keras, PyTorch, TensorFlow, Armadillo, `mlpack`, and so on. Despite their popularity, building ML models is not trivial as it requires a good understanding and background in data science. For this chapter, we will use scikit-learn.

scikit-learn

scikit-learn is a Python package written for the sole purpose of doing ML and is one of the most popular ML libraries used by data scientists. It has a rich collection of ML algorithms, extensive tutorials, good documentation, and most importantly excellent user community. For this introductory chapter, we will use scikit-learn for developing ML models in Python. Wherever applicable, we will use the scikit-learn `1.0.2` version, and we will look at separate ways of installing scikit-learn in the subsequent chapters.

Genomics big data

Genomics is the study of the function, structure, and evolution of genomes in living organisms. A genome is the blueprint of an organism that has a complete set of DNA, including genes and other intergenic regions. Genes are the basic components of DNA, and they play an important role in inheritance. The field of genomics got mainstream attention after the completion of human genome sequencing in 2003. The human genome project catapulted the field of genomics and it transformed medicine, giving birth to the modern biotechnology industry. Genomics got another push with the introduction of **Next-Generation Sequencing** (**NGS**) in the early 2000s, which enabled researchers and scientists to generate massive amounts of data, leading to scientific breakthroughs.

Genomic data has gained a lot of attention in the last decade because of the incredible progress it has made in precision genomics, genomic medicine, drug development, therapeutics, and so on. For example, since the first SARS-CoV-2 (the causative organism of COVID-19) genome was sequenced in December 2019, we have had around 3,035 SARS-CoV-2 genomes sequenced up to May 2022 (`https://nextstrain.org/ncov/gisaid/global/all-time`), which enabled companies such as Moderna to quickly develop boosters based on the sequence of the variants. These technological advancements in genome sequencing have enabled scientists to create and store that massive data to drive biomedical breakthroughs. For example, a lot of effort has been put into the generation of deep datasets such as **The Cancer Genome Atlas** (**TCGA**, 2014), the **Encyclopedia of DNA Elements** (**ENCODE**, 2012), **GenBank** from the **National Center for Biotechnology Information** (**NCBI, 1988**), and so on.

Big data is a term that is used to refer to datasets that are larger in size and complexity in terms of type compared to traditional data generation methods.

While genomic big data has enabled life sciences and biotechnology industries to make informed business decisions, this continuous generation of big data has put several constraints on the collection, acquisition, storage, analysis, and sharing of data. In addition, it is estimated that there will be between 2 and 40 **exabytes** (**EB**) of genomics data generated in the next decade (`https://www.ncbi.nlm.nih.gov/pmc/articles/PMC4494865/`).

But thanks to ML, we can now translate this genomic big data into actionable insights that can be applied for research and scientific innovation.

Supervised and unsupervised ML

The goal of ML is to develop and deploy computational algorithms that can automatically learn and improve from experience without human interference to perform a particular task. But how does it work? It does so by first "*learning*" knowledge from experience from the input data and using that knowledge to make predictions on unseen data. As such, the crux of ML is the **learning problem** in which machines learn from real-world data, improve from experience, extract patterns, construct models, and predict the outcomes of unseen data.

Depending on the type of data and the tasks to perform, ML algorithms can be broadly divided into supervised, semi-supervised, and unsupervised methods. Supervised methods learn patterns from examples with *labels* (for example, "diseased" or "not diseased") and are then used to predict future events or labels from unseen data (*Figure 3.1*). Unsupervised methods, in contrast, don't have the luxury of labels, and they rely on examples only to find patterns in the data and create groups (*Figure 3.1*). Semi-supervised methods combine both supervised and unsupervised methods, using patterns in the unlabeled data and using those to improve the prediction power of labels. Semi-supervised methods are not as popular as the other two approaches and so we will not discuss them in this chapter.

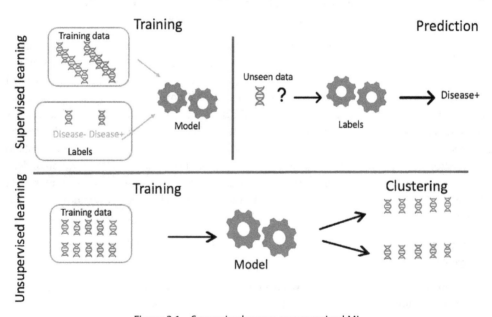

Figure 3.1 – Supervised versus unsupervised ML

Let's discuss these supervised and unsupervised learning methods with some examples here and in the latter part of the section.

Supervised ML

Supervised ML, as the name suggests, is a class of ML methods that rely on the *labels* provided during the training of a computer algorithm along with the *features* from the input data. In other words, it is a type of ML that learns a mapping from input X to a set of outcomes for quantitative or categorical values in the training set. This process of learning is called "modeling". The observed value of the variable X can be represented by a NxM dimensional matrix of elements where N is the number of observations (for example, number of genes or number of patients/samples) and M is the number of features (for example, **single nucleotide polymorphisms** (**SNPs**) or gene expressions values). Y is an N-dimensional vector of output variables assuming continuous or categorical values. *But how do the models know how to learn the mapping from the input to output or modeling* is a common question among beginners, but fortunately, we have a lot of information on this process. But before we go into deeper details of supervised ML, let's look at the different types of supervised ML methods.

Types of supervised ML

Supervised ML methods can be mainly divided into regression and classification based on the most common use cases. A regression model predicts a continuous value (*Figure 3.2*). Regression-based supervised ML methods employ parametric or non-parametric methods to map the associations of *features* to *lables* that are either quantitative or continuous (for example, the amount of gene expression). Several ML algorithms are currently available for performing regression tasks (*Figure 3.2*). On the other hand, in classification-based supervised ML methods, the output class is specified by a qualitative or discrete or categorical label (for example, type of gene— protein-coding or non-coding) with no explicit ordering (*Figure 3.2*). The different algorithms for classification tasks are shown in *Figure 3.2*. Classification models are divided into two different types: binary classification and multiclass classification. As the name suggests, binary classification models output a value from a class consisting of two values—for example, whether a particular DNA sequence is a "gene" or "not gene". Multiclass classification models, on the other hand, output a value from a class that contains more than two values—for example, a model that classifies either the **5'untranslated region (5'UTR)** or **coding sequence (CDS)** or 3'untranslated region (3'UTR):

	Regression	**Classification**
Outcome	Continuous value	Discrete/Categorical
Examples	Linear regression, Ridge regression, Lasso regression, Polynomial regression, and Bayesian regression	Logistic regression, K-nearest neighbor (KNN), Naïve Bayes, Support Vector Machine (SVM), decision trees, Random Forest, XGBoost

Figure 3.2 – Types of supervised ML

How do supervised ML methods work?

At the core of these supervised ML models are optimization methods, which are generally meant for minimizing a *loss* function. A loss function takes the predicted values and the values corresponding to the real data value and outputs how different they are. For example, in the case of linear regression models, the difference between the predicted methylation value and the original values of the labels' methylation for samples is considered as "loss". The lower the loss, the better the model. During model training, supervised methods try to learn the structure of the data to predict a value. Let's understand this further:

- The mapping function maps a predictor variable from a given sample to the response variable or labels. The response variable is also called the dependent variable. Then, the predictions are simply the output of the mapping function. The variables in the input are also called independent variables, explanatory variables, or features. The output of the mapping function is then compared to the original target value and loss is calculated. If the loss is high, then the model is modified until the loss is zero or minimal.

- The mapping functions also have parameters that help map input data to the predicted values. The optimization works on the parameters and tries to minimize the difference between the predicted output and the original value. This is the main idea of a cost function for regression, which is slightly different from a classification model, but the idea is the same.

- Finally, we need an evaluation method that informs of the predictions of a mapping function. Some of the well-known regression algorithms are **root-mean-square error** (**RMSE**), **mean absolute error** (**MAE**), R-Squared, and so on.

The key ingredients of a supervised ML algorithm can be summarized are as follows:

1. Define how to represent a mapping function $f(X)$ into a model for the machine to understand. This requires two things:

    I.   The first is to convert each input data or sample into a set of features or attributes that describe the purpose

    II.   The second is to pick a learning model (regression, in the case of continuous labels, and classification, in the case of discrete labels) that learns the system

2. Devise a *loss* function to optimize the difference between the predictions and the observed values. After converting the problem into a mapping problem, the step is choosing the right cost function that can calculate the difference between the predicted output and the original output.

3. Apply a mathematical optimization method to find the best parameter values in relation to the loss function. Train the chosen model iteratively until we find a model that best fits the data (loss is minimal) by tuning the parameters.

4. Apply an evaluation method that defines some metrics or scores for the prediction of the mapping function.

Let's next move on to unsupervised ML.

Unsupervised ML

While supervised ML methods aim to predict the output given the features and the labels in the input data, unsupervised ML methods aim to describe the associations in the data and uncover hidden patterns among a set of input variables so that they are meaningful (for example, low-expressed versus high-expressed genes). During this process, they can draw inferences from the datasets using their own rules. Since there are no labels in this dataset, these methods try to group samples based on similarity—for example, separating patients into different disease groups based on similar gene expression profiles. To do that, we first need to define a distance or similarity metric between a patient's expression profile and use that metric to differentiate patients that have one type of disease compared to the other patients or diseased vs non-diseased patients. There is no direct measure of success unless you have a label.

Types of unsupervised ML

Unsupervised ML methods can be broadly categorized into clustering and dimensionality reduction methods.

Clustering

Unsupervised ML learning methods solve problems by finding some useful structure in the data. **Clustering** is the most common unsupervised ML algorithm routinely used in genomic data analysis. Clustering is partitioning a set of observations into groups or clusters such that the pairwise dissimilarities between observations in the same group or cluster are smaller than those in other groups or clusters. After we discover this structure in the data in the form of clusters or groups, this structure can be used for tasks such as generating a useful summary of the input dataset or visualizing the structure of unknown sequences. If you have labels, then we can color the clusters with labels to see how well the clustering algorithm worked. The clustering of genomic sequences is one of the key applications in the field of genomics. The main challenge in clustering is the identification of groups/clusters and interpretation of the identified groups/clusters. There are methods such as the *elbow* method that can be used to determine the optimal number of groups or clusters.

Dimensionality reduction

The second popular unsupervised ML method is dimensionality reduction, where the goal is to reduce the number of variables in the data by obtaining a set of principal variables or components that explain the variance in the data. For example, in the case of a gene expression matrix across different patient samples, this might mean getting a new set of variables that cover the variation in sets of genes. **Principal component analysis** (**PCA**) is the most popular technique for reducing the dimensionality of highly complex data. Other popular methods are routinely used in genomics, such as **multidimensional scaling** (**MDS**) and **singular value decomposition** (**SVD**).

ML for genomics

Thanks to rapid advancements in NGS, genomics has shown tremendous growth in the last decade, which has led to an outpouring of massive sequence data. In addition to **whole-genome sequencing (WGS)**, other promising techniques have emerged, such as **whole-exome sequencing (WES)** to measure the expressed region of the genome, **whole-transcriptome sequencing (WTS)** or **RNA-sequencing (RNA-seq)** to measure mRNA expression, **ChIP-sequencing (ChIP-seq)** to identify transcription-factor binding sites, and **Ribo-sequencing (Ribo-seq)** to identify actively translating mRNAs for quantifying relative protein abundance, and so on. The challenge now is not "what to measure " but "how to analyze the data to extract meaningful data and turn those insights into applications". While the development of NGS technologies and the generation of massive data has provided opportunities for a new field called "bioinformatics" to grow significantly, it has also opened the door for the application of ML techniques with the goal of mining these large-scale biological genomic datasets.

ML for genomics is much needed as our understanding of genomes (including humans) is vastly incomplete. Uncovering discoveries can lead to the development of novel biological insights and can then be translated into genomic-based applications. The main areas where ML can help genomics include data characterization, pattern detection, correlation, classification, regression, cluster analysis, outlier analysis, and so on. However, the key to implementing ML in genomics is a clear understanding of ML workflow as genomics data requires significant domain expertise in every step of the process, right from data collection to model evaluation. With that goal in mind, let's look at the different components of an ML workflow next.

The basic workflow of ML in genomics

The basic process for an ML pipeline and the building of an ML model for a genomics dataset might be intimidating for a non-expert since it requires significant domain expertise in data collection, data cleaning, quality control of the genomic datasets, performing **exploratory data analysis (EDA)**, and so on. In addition, model tuning, validation, and deployment depend on a good understanding of biology and limitations and biases with data collection, methodology, and technology (*Figure 3.3*):

Figure 3.3 – A basic ML workflow for a typical genomic data analysis

Let's go into each of these components shown in *Figure 3.3*, from data collection to model monitoring, in detail now.

Data collection to preprocessing

The first step after project initiation and specification is data collection. Raw data can be generated from genomic sequencing from an experiment based on a hypothesis or extracted from public databases of various sources. After data collection, the next step is data preprocessing, which is the process of converting raw data to processed data to remove missing data, inconsistent data, and so on. Data preprocessing is a key step because low-quality data will affect the information extraction process, so removing incomplete or low-quality data or imputing missing data is important. Following the data preprocessing step is the data exploration and visualization step to understand more about the data. We looked at these first three steps (data collection, data preprocessing, EDA, and a little bit of feature extraction) as part of the genomic data analysis in the previous chapter and so we will not discuss these in this chapter.

In this chapter, we will briefly look at the rest of the steps—feature extraction, feature selection, train-test split, model training, model evaluation, and model tuning. In the concluding section of the book, we will investigate the topics related to model tuning, model deployment and monitoring.

Feature extraction and selection

Feature extraction and selection are considered the most important steps in ML as the performance of the model mainly depends on the features that you extract and include in the model training. Feature extraction, as the name suggests, is a process of creating features or attributes from datasets. It is the transformation of processed data into tabular data to represent a feature vector. The features can be numerical, categorical, ordinal, and so on. With feature selection, you select only relevant features that best fit the model based on statistical or hypothesis testing.

Train-test data splitting

Before you can feed the input data containing features and labels into various models such as linear/logistic, decision trees, Random Forest, SVM, Naïve Bayes, and so on for training the model, it is important to split the datasets into training and testing sets. This is because if we use the whole dataset for training, how can we test whether the model fit the data or not? The training set is meant for the model to learn and identify meaningful and generalized patterns during the training of the algorithm. The testing set is reserved for the evaluation of the final model after training. In many cases, it is recommended to split the data into three sets. In addition to training and test sets, we should split the data into a validation set for model selection and hyperparameter tuning. The data splitting could be constructed in such a way that it reflects real-world challenges and the size of the data.

Model training

Model training refers to when the training set is used in the optimization of the loss function to find parameters for function $f(X)$. Algorithms are trained using the training set, using a metrics such that the loss L determines how well the model predicts the output from an input in the case of supervised regression models. If the loss is minimal, consider that an optimal model.

Model evaluation

The optimized model can then be used to make predictions on a test set using several evaluation metrics such as **MSE**, MAE, RMSE, R-Squared, accuracy, and so on.

Model interpretability

Even though ML models make accurate predictions, they often can't explain their predictions in terms that humans can understand. The features used in the model training and used to conclude can be so numerous and their calculations so complex that is impossible to find out why the algorithm produced the answers. The ability to determine how an ML algorithm arrived at its conclusions is termed "model interpretability". Several model-agnostic interpretability methods such as **Local-Interpretable Model-agnostic Explanations (LIME)**, **SHapley Additive ExPlanations (SHAP)**, **Explanation Summary (ExSUM)**, and so on exist to help researchers interpret the results from the optimized model—for example, which features are important in the model—so that they can be experimentally validated in the wet lab before putting them into production.

Model deployment

This is where the optimized models are put into production to make predictions on real-world unseen data without labels and predictions from the model.

Model monitoring

This is the final step of the workflow where we monitor the model for its performance post-deployment.

Now that you have understood the detailed workflow for ML applications for genomics, let's dive deep into a real-world use case and see how it can be done using some Python and ML libraries.

An ML use case for genomics – Disease prediction

Let's illustrate the power of ML for genomic applications, starting with classification models, which are a subset of supervised ML methods where the goal is to classify the outcome into two (binary classification) or more (multiclass classification) classes based on the independent variables.

One of the popular use cases for genomics is outcome prediction. In this particular use case, we will try to predict if a patient has lung cancer or not based on gene expression. Before we start building the model and using that to make a prediction, let's try to understand how a typical ML disease prediction model work in this use case. It works by mapping the relationships between individual patients' sample gene expression values (features) and the target variable (Normal versus Tumor)—in other words, mapping the pattern of the features within the expression data to the target variable. In this example, we will use a supervised ML method to build a classification model from the expression data to predict the outcome. Each row of the data represents a patient sample that consists of gene expressions. We will use **logistic regression** to build a simple binary classification model for outcome prediction.

The workflow consists of the following steps:

1. **Data collection**: Where we download the data and load it into the system
2. **Data preprocessing**: Where we clean, normalize, and standardize the data if required

3. **EDA**: Where we visualize the data

4. **Data transformation**: Where we transform the data

5. **Data splitting**: Where we split the data into training and testing sets

6. **Model training**: Where we train the model using the training data

7. **Model evaluation**: Where we evaluate the data using the test data

Let's cover these steps in detail next.

Data collection

We will start our illustration of ML on a genomics problem using a real dataset from *BARRA:CuRDa*, a curated RNA-seq database for cancer research (https://sbcb.inf.ufrgs.br/barracurda). RNA-seq is one of the most important methods for inferring global gene expression levels in biological samples.

CuRDa is a repository containing 17 handpicked RNA-seq datasets, extensively curated from the **Gene Expression Omnibus** (**GEO**) database using rigorous filtering criteria. We will use the gene expression data of lung cancer samples from that repository (which we have already downloaded and provided it you in here - https://github.com/PacktPublishing/Deep-Learning-for-Genomics-/tree/main/Chapter03/lung), and we will try to predict normal versus tumor outcomes using the expression data. There are two data objectives that we need for this exercise, one for the gene expression values for each sample and the other for type (normal or tumor). This dataset is extremely small for real-world application, but it is very relevant for the genomics focus of this section, and small datasets take very little time to train. Please note that for ease of understanding we are showing the details of the individual steps of the whole process but in real life you will code all these steps in a single script as shown here (https://github.com/PacktPublishing/Deep-Learning-for-Genomics-/blob/main/Chapter03/Disease_prediction_LR_CuRDa.py)

Here are the steps.

1. First, we will load the data into Pandas using the read_csv method and concatenate them into a single dataframe:

```
import pandas as pd
lung1 = pd.read_csv("lung/GSE87340.csv.zip")
lung2 = pd.read_csv("lung/GSE60052.csv.zip")
lung3 = pd.read_csv("lung/GSE40419.csv.zip")
lung4 = pd.read_csv("lung/GSE37764.csv.zip")
lung_1_4 = pd.concat([lung1, lung2, lung3, lung4])
```

> **Note**
>
> Only the lung sample data was collected and downloaded as a CSV file. We will use that for all ML model-building purposes.

2. Let's print the first 5 rows and 10 columns using Pandas' head method:
    `lung_1_4.iloc[:,0:10].head():`

	ID	class	ENSG00000000003	ENSG00000000005	ENSG00000000419	ENSG00000000457	ENSG00000000460	ENSG00000000938	ENSG00000000971	ENSG00000001036
0	SRR4296063	Normal	10.728260	4.668142	10.278195	10.184036	8.215333	11.310861	13.178872	11.469473
1	SRR4296064	Tumor	11.332606	2.329988	10.127734	10.167900	8.174060	10.399611	13.208972	11.510862
2	SRR4296065	Normal	9.951182	4.264426	10.288874	10.093258	8.011385	11.814572	14.038661	11.651766
3	SRR4296066	Tumor	12.185680	2.798643	10.178582	10.401606	8.902321	10.294009	13.170466	11.546855
4	SRR4296067	Normal	9.875179	2.922071	10.444479	10.435843	8.692961	12.604934	13.538341	11.733252

Figure 3.4 – First 5 rows and 10 columns using Pandas' head method

What do you see? We see that there are column names such as ID, class, ENSG00000000003, and so on. The ID indicates the SRA ID from where the sample is coming from, the class here indicates whether the sample is classified as normal or a tumor, and the rest of the columns represent the gene expression values of the sample.

Data preprocessing

Generally, raw data needs to be preprocessed before we start training. This means converting raw data to something clean by removing any outliers, null values, missing values, and correlations between the predictor variables. As many ML algorithms are sensitive to these, we should deal with this in the first step. In many cases, data consists of missing values, and there are two ways to deal with the missing data—remove it or impute it.

We will see how to do this in practice using the Pandas package. First, look at the amount of missing data in the data by running the isna() method on the DataFrame and then taking a sum of that using sum() method, like so:

```
lung_1_4.isna().sum()
ID 0
class 0
ENSG00000000003 0
ENSG00000000005 0
ENSG00000000419 0
. .
ENSG00000285990 0
ENSG00000285991 0
ENSG00000285992 0
ENSG00000285993 0
ENSG00000285994 0
```

As you see, none of the columns has any missing values, which is good for us since we don't have to deal with missing values. Since there are a lot of columns in this dataset and the preceding output doesn't really indicate if there are any missing values or not, let's take a sum of the missing columns for all columns now:

```
lung_1_4.isna().sum().sum()
0
```

The preceding result also indicates that there are no missing values for any columns. We will move on to the next step, then.

EDA

The first step around any data-related analysis is to start by doing EDA. This can be done by looking at the distributions of the data itself. Since EDA is quite crucial for ML, we will try to visualize the data to understand the data structure in general.

Here are the steps:

1. Let's first start by plotting the distribution of samples corresponding to each lung cancer type:

    ```
    df = lung_1_4['class'].value_counts().reset_index()
    ```

    What have we done here? We first created a DataFrame of the class column, then calculated the number of rows corresponding to each class, and then reset the index to make it easy for plotting.

2. Next, we will visualize the classes on a bar plot for easy visualization of class distribution for this dataset using the Seaborn library:

```
import seaborn as sns
import matplotlib.pyplot as plt
sns.barplot(x = "class", y = "index", data=df)
plt.xlabel("Number of samples")
plt.ylabel("Class")
```

Let's visualize the plot now:

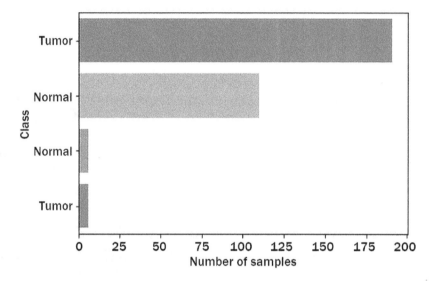

Figure 3.5 – Plotting of lung cancer classes and their values in a bar plot

This bar plot in *Figure 3.5* shows how many samples fall into each of Normal and Tumor (lung cancer) types. Oh, no. We have a problem now. As you can see, there are two types of samples, both of which are classified as Normal and the same for Tumor. What be the reason?

3. Let's look at the different classes closely and see what's going on:

```
set(lung_1_4['class'])
{' Normal', ' Tumor', 'Normal', 'Tumor'}
```

Do you see anything weird here? If you look closely, we notice that there is an extra space in front of the first and second classes, because of which we get two extra classes, even though the total number of classes should be only two. This is quite common for public datasets, and this is where EDA will come in handy. Fortunately, this is an easy fix.

4. Let's rename those right away using the following `replace` method:

```
lung_1_4['class'] = lung_1_4['class'].replace(' Normal',
'Normal')
lung_1_4['class'] = lung_1_4['class'].replace(' Tumor',
'Tumor')
```

5. Now that we have fixed the issue, let's replot using the same code as before and see the class distribution:

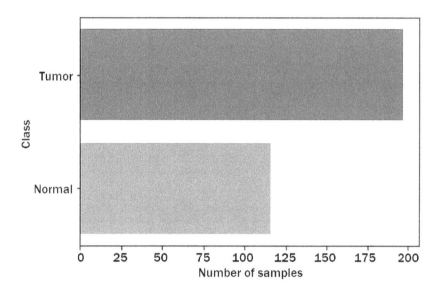

Figure 3.6 – Plotting of lung cancer classes and their values in a bar plot after fixing the labels

Now, the plot looks much better and is kind of expected. There are some differences in the number of samples that belong to the `Normal` class (197) compared to the `Tumor` class (116), which is expected. This class imbalance will create problems during model predictions, but there are advanced techniques available for addressing class imbalance. However, it is beyond the scope of this tutorial to discuss those techniques.

Data transformation

The next step after data preprocessing and EDA is data transformation. This is an important step because many times, each variable/feature has a different magnitude. So, it is a good practice to scale the data that might come from how the experiments are performed and the potential problems that might occur during data collection. Any systematic differences between samples must be corrected before proceeding to the next step. In this step, we will check if there are such differences using box plots.

Here are the steps:

1. First, we will restrict our dataset to the first 10 columns since it is challenging to visualize all the columns at once in a single boxplot. Then, we convert the data from wide format to long format using the `melt` method in Pandas:

```
lung_1_4_m = pd.melt(lung_1_4.iloc[:,1:12], id_vars = "class")
```

2. Next, we will use the Seaborn visualization library to look at the distribution of expression across selected samples:

```
ax = sns.boxplot(x = "variable" , y = "value", data = lung_1_4_m, hue = "class")
ax.set_xticklabels(ax.get_xticklabels(), rotation=90)
plt.xlabel("Genes")
plt.ylabel("Expression")
```

Now, let's visualize the boxplot:

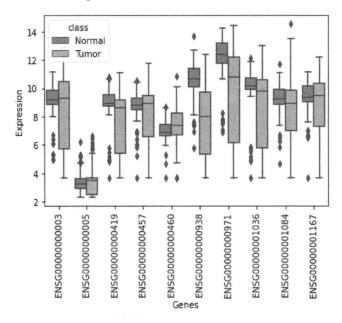

Figure 3.7 – Boxplot showing the gene expression value distribution for 10 samples

As shown in *Figure 3.7*, each sample has a somewhat similar distribution of gene expression values except for the first few samples (compare the medians). In addition, the expression values are already normalized and there is no need to normalize this further. So, let's proceed without normalizing these samples.

Data splitting

Once the data is transformed and scaled we are now ready to train the model, but before that, we must split the data into train and test datasets. The reason for splitting into train and test is to have a training set for training the model and keep an independent dataset (test set) for model evaluation.

> **Note**
>
> For this simple exercise, we will not split it into three different datasets (train, validate, and test).

For this purpose, we will use Scikit-learn's `train_test_split` function, which can split the data into train and test based on the split ratio. In this case, we will split the train and test datasets in the ratio of 75:25 because we want to use more data for training than for testing, but in the real world, we will have ratios something like 60:40 training : testing.

Data splitting is a multi-step process, so let's understand this in detail next:

1. Drop the `ID` and `class` columns in the dataset, and convert it to a NumPy ndarray, a multidimensional container of items of the same type and size:

    ```
    x_data = lung_1_4.drop(['class', 'ID'], axis = 1).values
    ```

    Similarly, we will create a NumPy ndarray for the labels from the subset data:

    ```
    y_data = lung_1_4['class'].values
    ```

2. Next, we must first convert the categorical data in the `type` column to numbers. Ordinal encoding and one-hot encoding are the two most popular techniques to convert categorical data to numbers. In this simple tutorial, we will use the ordinal encoding method. Let's create a variable class to create a unique list of the different classes in the target:

    ```
    classes = lung_1_4['class'].unique().tolist()
    classes
    ['Normal', 'Tumor']
    ```

3. Next, convert the classes into ordinals using a custom function:

    ```
    import numpy as np
    func = lambda x: classes.index(x)
    y_data = np.asarray([func(i) for i in y_data], dtype = "float32")
    ```

    Printing the y_data array (not shown here) shows the different classes converted to ordinals first and then to floats. Here, 0 represents the Normal class, while 1 represents the Tumor class.

4. Now, we are ready to split the data into training and testing. For this, we will use Scikit-learn's `train_test_split` function:

```
from sklearn.linear_model import LogisticRegression
from sklearn.model_selection import train_test_split
X_train, X_test, y_train, y_test = train_test_split(x_
data, y_data, random_state = 42, test_size=0.25, stratify
= y_data)
X_train.shape, X_test.shape, y_train.shape, y_test.shape
((234, 58735), (79, 58735), (234,), (79,))
```

The `train_test_split` function takes multiple arguments to split the features and labels into four different outputs. For this example, we are specifying a train : test split ratio of 75:25, and also, we are making sure that we will stratify (to make sure that the ratios of all the classes - Normal and Tumor are maintained equally between train and test datasets) the data first before splitting.

Model training

Finally, the exciting part of the exercise is training the model using the training set that we split in the previous step. For this simple exercise, we will use **logistic regression**. Logistic regression by default classifies data into two categories, which is exactly what we want here. Let us run the Logistic regression with our training dataset. Again, we will use scikit-learn's linear model method for running Logistic regression on the training data:

```
model_lung1 = LogisticRegression()
model_lung1.fit(X_train, y_train)
```

Here, we are first instantiating an object using the `LogisticRegression` function and using that to fit the training data consisting of features and labels.

Model evaluation

Now that model has been trained, let's run the model on one sample of the test data. Running the following command outputs the label as part of the prediction:

```
pred = model_lung1.predict(X_test[12].reshape(1,-1))
pred
array(['Tumor'], dtype=object)
```

Here, we are using the trained model (`model_lung`) to predict on one of the samples in the test data (12). The prediction returns a `normal` sample. As a manual validation, you can check and see if the prediction here matches the actual label in the test data. You can also do predictions for all samples in the test data:

```
all_pred_lung= model_lung1predict(X_test)
```

The output here (not shown) indicates the predicted classes for each of the samples in the test data. Now, you can check to see if the class that is predicted by this algorithm is the same as the original class or not.

So far, we looked at model prediction on a single sample, but can we trust the model just based on a single prediction? We will have to check the model predictions on all the test data and compare them with the original labels in `y_test` data. To assess the performance of our model, we must first define some metrics. There are several metrics that we can use, and it's beyond the scope of this chapter to go over all of them. We will use two popular evaluation metrics here:

1. The first is `accuracy_score`, which calculates the accuracy of the model. Accuracy is defined as the number of true positives and true negatives divided by all predictions. Have a look at the following code snippet:

   ```
   model_lung1score(X_test, y_test)
   0.9620253164556962
   ```

   As can be seen from the output, the model has a very score of 96% without doing any optimization (such as cross-validation, model selection, hyperparameter tuning, and so on). This is not bad for this dataset.

2. Let's run a confusion matrix to indicate how well the model did in terms of true positives, true negatives, false positives, and false negatives:

   ```
   from sklearn.metrics import confusion_matrix ,
   ConfusionMatrixDisplay, classification_report
   cm = confusion_matrix(y_test, all_pred_lung)
   disp = ConfusionMatrixDisplay(confusion_matrix=cm,
   display_labels = ["Normal", 'Tumor'])
   disp.plot()
   plt.show()
   ```

Let's look at the following screenshot as the confusion matrix of the Logistic regression model:

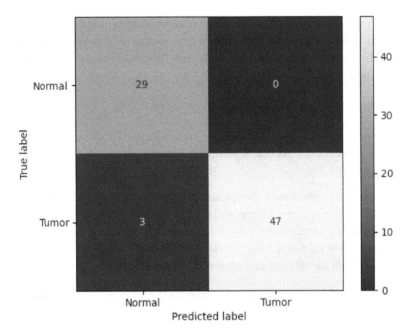

Figure 3.8 – Confusion matrix of the Logistic regression model

Overall, the model did really well classifying the samples correctly for Normal and Tumor. There are a few instances where the model misclassified the samples. For example, there is one sample that the model predicted as Tumor but that is actually Normal (false positives), and there are three samples that the model predicted as Normal but are Tumor samples (false negatives). Please note that the cost of misclassifying a sample is high for false negative samples compared to false positive ones because we don't want to miss any patient that has a tumor.

3. The second popular metric is classification_report. Let's use classification_report from Scikit-learn to print classification metrics for our model:

classification_report(y_test, all_pred_lung)

Let's look at the following results of the classification report of the lung cancer samples:

	precision	recall	f1-score	support
Normal	0.91	1	0.95	29
Tumor	1	0.94	0.97	50
accuracy			0.96	79
Macro-avg	0.95	0.97	0.96	79
Weighted-avg	0.97	0.96	0.96	79

Figure 3.9 – Classification report of the lung cancer samples

The most important component of the `classification_report` metric for classification models is the F1 score, which is the harmonic mean of precision and recall. Here, the F1 score is very good and is close to 1 for all the classes, which indicates that the model has done a good job of predicting the correct class of the sample.

ML challenges in genomics

ML is the backbone of the large-scale analysis of genomic data. ML algorithms can be used to mine biological insights from genomics big data and discover predictable patterns that may be hard to extract by experts. However, there are a few challenges the current ML algorithms face in the analysis of genomic data:

- Although the amount of data coming out of biological systems and genome sequencing is huge and ever-growing, integrating these diverse datasets from multiple sources, platforms, and technologies into ML algorithms is not trivial.

- Because of this huge variation in trained data, models tend to overfit and they generalize very poorly on new data that is different from the training data. We can use methods such as L1 and L2 regularization to address this poor generalization, which we will see in future chapters.

- The nature of ML models, which are mostly "black boxes", may bring new challenges to biological applications in particular models to predict diseases. Despite several ML interpretability tools, it is hard to interpret the model predictions, especially from a biological point of view, restricting their application in genomics and other biological domains.

- An ML model needs hand-crafted features, which require domain knowledge in biology and genomics. This is a significant problem for non-experts in genomics.

These are a few of the ML challenges in genomics, and in the next chapter, we will look at sophisticated techniques such as **deep learning** (**DL**), which can address some of these challenges in genomics. Before we end this chapter, let's briefly summarize what we have learned so far in this chapter.

Summary

This chapter started with what ML is and how ML algorithms can help genomic applications through their inherent nature of uncovering hidden patterns in the dataset, automating human tasks, and making predictions on unseen data. We looked at the several types of ML algorithms—namely, supervised and unsupervised methods—and understood the main steps in ML methods. Then, we understood the ML workflow for genomic applications.

In the second half of the chapter, we spent quite a bit of time understanding the different steps in ML and what is involved in each step of the workflow. We also introduced the most popular `Python` packages `Pandas` and scikit-learn to work on the ML workflow. Finally, we worked on a real-world application of ML on a genomic dataset for identifying the disease state of cancer patients.

This chapter and the preceding chapters are meant for a quick primer on ML for genomics, and with this knowledge and understanding of fundamentals, in the next chapter onward, we will dive into DL for genomics and understand the different algorithms for analyzing large and complex genomic datasets for unraveling meaningful biological insights.

Part 2 –
Deep Learning for Genomic
Applications

The primary goal of this part is to transition you from machine learning to deep learning and introduce you to different deep learning technologies specific to genomic applications, using real-life examples to transform raw genomics data into biological insights.

This section comprises the following chapters:

- *Chapter 4, Deep Learning for Genomics*
- *Chapter 5, Introducing Convolutional Neural Networks for Genomics*
- *Chapter 6, Recurrent Neural Networks in Genomics*
- *Chapter 7, Unsupervised Deep Learning with Autoencoders*
- *Chapter 8, GANs for Improving Models in Genomics*

Deep Learning for Genomics

Recently, there has been a rapid increase in interest in genomics-based applications in the biomedical, pharmaceutical, and therapeutics industries. **Machine learning (ML)**, with its sophisticated mathematical and data analysis techniques, coupled with advances in **next-generation sequencing (NGS)** have played a huge role in this rapid rise. As most genomic companies and other research organizations started to produce genomic data to keep themselves ahead of the curve, the ability to extract novel biological insights and build predictive models from this ever-growing data has proved to be a challenge for ML because it relied on hand-crafted features for model training and predictions as we saw in the previous two chapters. Translating this massive genomic data from an incomprehensible resource into meaningful insights automatically and intuitively requires more expressive ML models and algorithms.

Deep learning (DL), a subcategory of ML that can extract features automatically and automate complex tasks involving feature learning and non-linear activations, has revolutionized several research fields and industries in the past few years. DL uses **artificial neural networks (ANNs)** to mimic the ability of a human brain to learn novel patterns through training on tasks and allows machines to solve complex problems and learn iteratively without human intervention. DL is becoming the method of choice for many genomic modeling tasks because it generates accurate predictions in specific tasks compared to state-of-the-art techniques such as ML and bioinformatics methods. Advances in DL algorithms were made possible through the development of high-level DL APIs such as PyTorch and Keras and, most importantly, the availability of hardware (**graphical processing units (GPUs)** and **TensorFlow processing units (TPUs)**) that power their training in a few lines of code and the generation of large-scale data using **next-generation sequencing (NGS)**.

This chapter will introduce you to the fundamentals of DL, diverse types of DL algorithms, Python libraries for DL, and real-world applications of DL for solving some key challenges in the field of genomics. By the end of this chapter, you will understand how DL works, be familiar with ANNs, and understand how DL algorithms, along with Python libraries, can be used to build DL models and make predictions on genomic big data.

As such, in this chapter, we are going to cover the following main topics:

- Understanding what deep learning is and how it works
- The anatomy of deep neural networks (DNN)

- DNNs for genomics
- Introducing deep learning algorithms and Python libraries

Understanding what deep learning is and how it works

In the past few years, ML has been the go-to tool for academic research and industries since ML made it possible to learn complicated functions and patterns from highly complex data without human intervention. As early as 1980, theoretical results such as **Universal Approximation Theorem** seemed to indicate that it may be possible for a neural network to learn any function that existed in a dataset. This is a powerful approach because there are several problems in the real world that traditional methods cannot solve. This led to the birth of DL. Even though DL has been around for about a decade now, it has gotten mainstream attention recently. So, why didn't DL take off until recently? This can be mainly attributed to the lack of DL frameworks, big data, and efficient hardware to build complex DL models until recently. It's only been possible to use DL to produce meaningful empirical results due to the introduction of cheaper and faster hardware such as GPUs and TPUs, which has enabled parallel computing, along with the availability of sophisticated algorithms and big data.

Neural network definition

Neural networks are a set of algorithms that are inspired by biological neural networks in the human brain to interpret large-scale complex data. By definition, neural networks provide multiple levels of representation of the data. They use a hierarchy of multiple layers where each algorithm in the hierarchy uses a non-linear transformation function for the input to predict the output. The iteration process continues until the predicted output reaches a desired level of accuracy. DL is extremely useful for those disciplines that collect, analyze, and interpret large amounts of data and hence is a great fit for genomics. We will discuss several applications of genomics later in this chapter but for now, let's understand the anatomy of neural networks.

Anatomy of deep neural networks

Neural networks are a collection of neurons that are interconnected with each other through various layers that can learn the mapping between the inputs and the outputs with the provided training data. Neural networks model complex patterns in datasets where traditional ML algorithms fail. They use multiple hidden layers and non-linear activation functions. The concept of a neural network was put first forth by Warren McCullough and Walter Pitts in 1944, and they describe a neural network as the collection of connected nodes (`https://www.cambridge.org/core/journals/journal-of-symbolic-logic/article/abs/warren-s-mcculloch-and-walter-pitts-a-logical-calculus-of-the-ideas-immanent-in-nervous-activity-bulletin-of-mathematical-biophysics-vol-5-1943-p-p-115133/7DFDC43EC1E5BD05E9DA85E1C41A01BD`). Here, nodes represent artificial neurons, which are the functional units of neural networks which we will discuss in the following section.

The fundamental unit of neural networks is the **artificial neuron** (**AN**), which mimics the biological neuron in the human brain. Before we discuss ANs, let's quickly refresh our memories of what a biological neuron is (*Figure 4.1*):

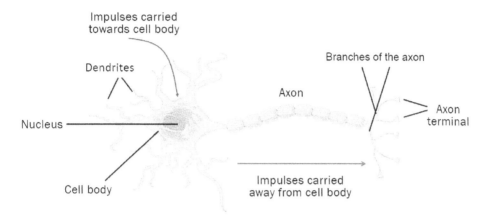

Figure 4.1 – Pictorial representation of a biological neuron

Note

The preceding figure was partly generated using Servier Medical Art, provided by Servier, licensed under a Creative Commons Attribution 3.0 unported license.

A biological neuron is a cell in the brain that transmit signals to other cells, such as nerve cells, muscle cells, gland cells, and so on:

- A typical neuron consists of a cell body, an axon, and dendrites. The cell body of a biological neuron consists of a nucleus and cytoplasm.

- The axon of the neuron extends from the cell body and gives rise to many smaller branches, called dendrites, before ending at nerve terminals.

- Dendrites receive signs from other neurons. The point of contact to communicate between neurons is called a synapse.

- When a neuron sends excitatory signals to other neurons, this signal is added to all other inputs of that neuron. If it exceeds a particular, threshold then the target neuron will fire and pass the signal forward.

Neurons represent a single cell in a biological neural network. The same concept applies to ANN, which simulates the basic functions of a biological neuron. However, some key differences exist between a biological neuron and an AN with regards to the size, topology, speed, signals, fault tolerance, and so on. We will briefly touch on some of these key differences throughout this chapter.

The basic architecture of a **DNN** represents abstractions of AN without all the biological complexities. They can learn from known examples and apply all those learnings to unseen data. In a DNN, the ANs in several layers receive input, process those inputs, and generate a result. They are typically aggregated into three layers: the **input layer**, the **hidden layer**, and the **output layer**. As shown in *Figure 4.2*, a typical DNN receives a set of features as input, passes them through one or more multiple hidden layers as part of the learning process, and ultimately makes the predictions in the output layer, which represents the combined input from all the neurons:

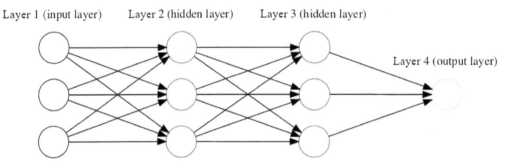

Figure 4.2 – Example of an AN or DNN

Let's understand each of the three layers of a DNN.

Input layer

The input layer of a neural network receives the data in several forms, such as images, text, audio, DNA sequences, and so on, and feeds it into the hidden layer. Like ML, input is the set of features that are provided to the model, which are then used for training the model and learning purposes. For example, in the case of a disease prediction model, the set of features for inputs can be the length of the gene, GC content, kmer frequency, expression value, and so on for each DNA sequence In advanced DNN architectures that will learn in future chapters, the features can be one hot encoded matrix.

Hidden layer

The hidden layers are what make neural networks different from other algorithms. These are layers between the input and output layers that are involved in computations. The inputs get aggregated using an activation function and then directed as output in this layer. There can be one or more hidden layers that are interconnected with each other through nodes; they are responsible for searching for different hidden features in the data. For example, in the case of disease prediction, the first layer might be responsible for higher-level features and the later layers perform more complicated tasks.

> **Note**
>
> It is quite common to have multiple nodes in the input and hidden layers and one or more output layer, depending on the type of algorithm. DNNs generally consist of many hidden layers. The number of ANs and hidden layers depends on the problem that you are solving and how big the dataset is. The number of nodes and layers in the node is an hyperparameter that can tune which you will learn in future chapters.

Output layer

The output layer is the final layer and is used to derive the output results based on the performance of the input layers and the hidden layer. It takes inputs from hidden layers, applies an non-linear (activation) function, and generates a final prediction (continuous or discrete), depending on the algorithm. The final output layer may be a single node or multiple nodes, and it depends on the problem and how the model is built (we will see some versions of these later in this chapter). In a way, the output layer is the most important as the model predictions are generated here.

Key concepts of DNNs

In addition to the three different layers that we just discussed, there are some key concepts of DNNs that we should all know as highlighted in (*Figure 4.3*).

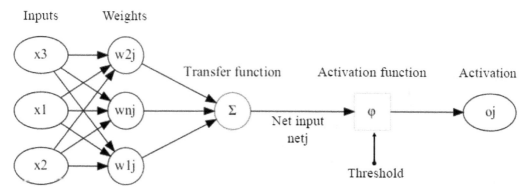

Figure 4.3 – Key concepts of DNNs

Let's go through each of the key components that make a neural network so powerful for solving complex problems (*Figure 4.3*):.

Weights

Weights represent the importance of the features in the input, as shown in *Figure 4.3*. Initially, these weights are randomly provided to the network and learned during model training in the backpropagation step. The higher the weight, the more important the feature is and its contribution to learning. For example, in the case of disease prediction, a lower weight is given to the gene length compared to GC content. The input values are multiplied by weights, along with an additional value (the bias) during model training. Weights are learned during model training.

Bias

Bias is one of the key components of the neural network. It is a value not shown in the above image in*Figure 4.3* that's added to the weighted sum of the inputs and the weights before it's fed to the activation function. The role of bias terms is to shift the value generated by the activation function. You can think of bias as a constant in a linear function. Bias terms are generally learned by the model during training.

Transfer function

The transfer function combines multiple input features into one aggregated value so that the activation function can be applied to it. It does this by simply taking a sum of all the inputs.

Activation function

The activation function is the key component of a neural network as it introduces the non-linearity in the architecture. By definition, they transform the summed weighted input from the input node (the output from the transfer function) into an output value to be passed on to the next layer in the architecture (*Figure 4.3*). Activation functions are unique components of any DL architecture. They help convert a linear relationship into a nonlinear relationship, which is the key to solving so many of the problems that cannot be typically solved by ML. Without the activation function, the network would be just a linear combination of input values. Activation functions will decide if an input to the network is important and should be passed on further or not. This is where DL is different from traditional ML; for example, without the non-linear activation functions, which is the case with ML models, the neural networks behave just like a linear regression function. The activation function helps the DNNs keep the most useful information and filters all irrelevant data points.

Since activation functions are very important, let's spend some time understanding the different activation functions and where they can be applied. Several types of activation functions are available. These will be described in the following subsections.

Binary step

The binary step function is the simplest of all the activation functions. The transformed value from the input node from the activation function is compared to the threshold. Depending on the threshold, the neuron will either be activated or deactivated. In this example (*Figure 4.4*), the threshold is 0 and so it return '0' if the input is less than 0 otherwise it returns 1.

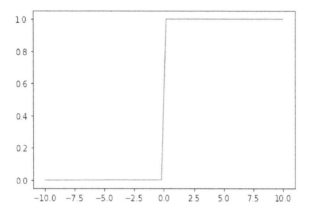

Figure 4.4 – Binary step activation function

Even though the binary step function can be implemented easily, it is not a good activation function for multi-class classification problems, where the goal is to predict one of many classes. In addition, the gradient of the binary step activation function is zero, which means it does not allow the backpropagation algorithm to adjust the weights.

ReLu

Rectified Linear Unit (ReLu) is the most popular activation function. It is simple to understand and implement. If the aggregation function is less than 0, neurons will be deactivated; otherwise, they get activated (*Figure 4.5*). Since only a few neurons are activated at a time, ReLu is more computationally efficient compared to other functions:

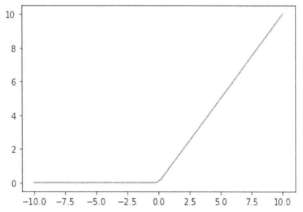

Figure 4.5 – ReLu activation function

Most importantly, ReLu help accelerates the convergence of gradient descent of the loss function toward the global minimum because of its non-linear and non-saturating property. It is also computationally efficient compared to other activation functions because only a few neurons are activated at a time.

Leaky ReLU

Leaky ReLU was introduced to address the dying ReLU problem, where the negative values make the gradient value zero and cause some dead neurons that never get activated. Leaky ReLU is a variant of ReLU.

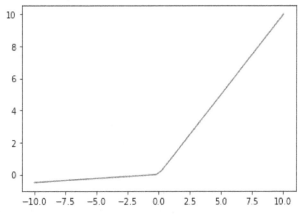

Figure 4.6 – Leaky ReLu activation function

It takes a small positive slope (0.05) for negative values instead of 0 and because of this, the gradient comes out to be a positive value, so there is no problem with dead neurons (*Figure 4.6*).

Sigmoid

The sigmoid activation function squeezes the output between 0 and 1. Because of this property, this activation function is good for predicting the probabilities of the fixed output range between 0 and 1. Unlike the binary step function, the sigmoid activation function is continuously differentiable and provides a smoother gradient during backpropagation, as shown here:

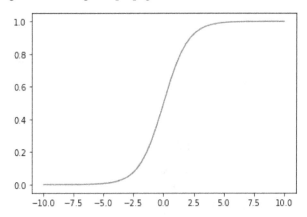

Figure 4.7 – Sigmoid activation function

However, it suffers from the vanishing gradient problem, where the network stops learning during model training.

Tanh

The tanh activation function is like the sigmoid activation function and has the same S-shape, but it squashes real-valued numbers between -1 and 1 (*Figure 4.8*). So, unlike sigmoid, the output of the tanh function is centered around 0 and it's easy to map the output as strongly negative (-1), neutral (0), or strongly positive (+1) as shown in the following figure:

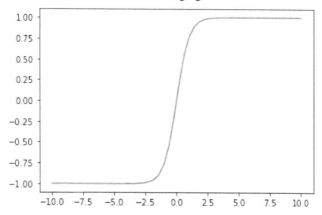

Figure 4.8 – Tanh activation function

This also suffers from the vanishing gradient problem.

Softmax

The softmax activation function is generally used in the output layer in the neural network for multi-class classification problems. It calculates the probability distribution of each target class over all the different possible target classes (*Figure 4.9*):

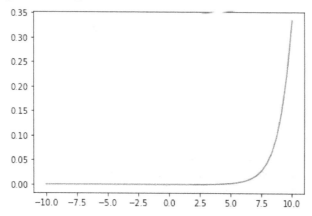

Figure 4.9 – Softmax activation function

For example, if there are three different classes, this shows the relative probabilities of the three different classes, which, when combined, equals 1.

Loss functions

In the case of supervised learning, for a given set of features in the output, we have a corresponding label in the output. During training, the algorithm learns the parameters (weights and biases) of the model so that it can accurately predict or classify the output. But how does the algorithm know when to adjust the parameters? This is where loss functions come in useful. A loss function tells the algorithm if the chosen model parameters are correct or not, and if not, it adjusts the parameters using gradient descent and backpropagation to make accurate predictions (we will see what gradient descent and backpropagation do later in this chapter).

In the case of classification algorithms, we use *cross-entropy loss*, which is the difference in the probability distributions of the classes. In the case of regression, we use the **mean squared error** (**MSE**),which is the average squared differences between the predicted values and the actual value. There are plenty of other loss functions out there but these two are the most popular loss functions for classification and regression, respectively.

Forward propagation

Forward propagation is the method through which neural networks make predictions. This network uses multiple layers (input, hidden, and output) to make predictions. For example, in this simple network, a single pass of forwarding propagation looks similar to what's shown in *Figure 4.10*:

Layer 1 (input layer)   Layer 2 (hidden layer)   Layer 3 (output layer)

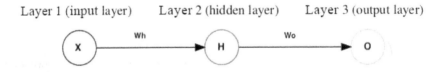

Figure 4.10 – A simplistic schematic of a forward propagation idea

Here, the input data is passed to a single hidden layer with an activation function such as ReLu to make the predictions in the output layer.

Backpropagation

Backpropagation is the opposite of forward propagation and is the most common algorithm for neural networks. It is the process of propagating the errors back into the network to update the weights at each node in the network so that they cause the original output to be closer to the target output, thereby lowering the error overall (*Figure 4.11*). It works by calculating the loss in the output layer by comparing the predictions with the observed values. The derivative concerning the weight is then calculated using the chain rule and then updates the weights, as shown in *Figure 4.11*:

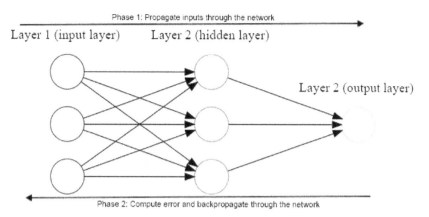

Figure 4.11 – Backpropagation concept in a nutshell

Briefly, this process can be summarized as follows:

1. Take a batch of data from the training dataset.

2. Perform forward propagation to calculate the loss.

3. Calculate the gradients of the loss with respects to weights and biases in the network by applying the chain rule (https://en.wikipedia.org/wiki/Chain_rule).

4. Use the gradients to update the weights of the network.

If we update the weight of each node iteratively, we will have a series of weights that generate good predictions. This is the main idea of backpropagation.

Gradient descent

Gradient descent is an optimization algorithm used to reduce the loss of a predictive model with regards to the train dataset.. The goal of a gradient descent algorithm is to find the local minimum of the loss function. We calculate gradients of the loss function by taking the derivative of the loss function and then iteratively going in the opposite direction of the derivative. The training data helps models train over time using the loss function and gradient descent optimization algorithm. This happens until the loss function is zero or closer to zero, indicating that the smallest possible error has been achieved and that the parameters have been updated. These models, which have been optimized for accuracy, can then be used for prediction and classification purposes for real-world applications.

Regularization

One of the key concepts in DL is to avoid biases and variance in the model. Among them, overfitting is the most important one. Regularization is a set of strategies used in DL to reduce the overfitting of the model and improve model predictions. Most models perform well after being trained on a specific subset of data but often, they fail on real-world data, which means they fail to generalize well. Regularization strategies aim to address overfitting and keep the training error as low as possible.

There are three types of regularization techniques. Let's take a look.

Lasso

In this method, the coefficients of the network are shrunk to 0 and because of that, it is suitable for variable selection.

Ridge

In this method, the coefficients of the network shrink to smaller values (but not 0).

Elastic Net

This method combines Lasso and Ridge and is a tradeoff between both methods.

There are several methods to combat overfitting in DL. Let's take a look.

Data augmentation

Data augmentation is an interesting regularization technique that can solve overfitting. The goal of data augmentation is to generate new training data based on a given original dataset and it provides a cheaper alternative to increase the amount of input data. This technique is very popular in **computer vision (CV)** and **natural language processing (NLP)**.

Dropout

This is another regularization method and among several regularization methods, dropout is the most popular. Dropout regularization is the process where the randomly selected neurons are dropped during the model training to prevent overfitting. It is a regularization method where it penalizes the nodes that have large weights.

We will see all these key concepts in action in the next section and other upcoming chapters.

An example of how neural networks work

Now that you've looked at the key components of a neural network, let's try and understand how a typical neural network works using a real-world example of *DNA methylation prediction*. **DNA methylation** is the process of adding a methyl group to a C5 position of the cytosine of a DNA sequence, resulting in 5-methylcytosine. DNA methylation is a key epigenetic mechanism that is involved in regulating gene expression. So, obtaining a precision prediction of DNA methylation is key in genomics.

For this example, we will extract 400 bps from multiple DNA sequences centered at the assayed methylation site (5-methylcytosine) and calculate GC content (counts of Gs and Cs divided by the length of the DNA sequence):

DNA Sample	GC
DNA1	0.6
DNA2	0.5
DNA3	0.4
.....	.....

Figure 4.12 – DNA sample and GC content

The preceding table shows the DNA sample ID and the extracted feature (**GC**) from each DNA sequence. Now, we can add the output to the table, which is methylation levels represented as a methylation ratio, ranging from 0 to 1 (*Figure 4.13*). The basic measurement used to quantify methylation is the methylation ratio, which is the log ratio of intensities observed in the treated sample compared to the control samples. 0 represents no methylation, while higher methylation values represent more methylation:

DNA Sample	GC	Methylation Value
DNA1	0.6	0.1
DNA2	0.5	0.4
DNA3	0.4	0.5
.....	.....	.....

Figure 4.13 – DNA sample, GC content, and methylation values

To come up with a model to predict methylation, what we do normally is fit a straight line through the data (GC content versus methylation value). But as we know, the straight line goes to the negative of the Y-axis and the methylation value cannot be negative. This means we must bend the line near 0 on the Y-axis. This is called non-linearity, and this is what the activation function in a neural network does (*Figure 4.14*):

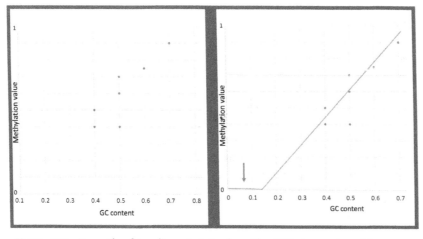

Figure 4.14 – Example of non-linearity introduced by ANNs to a real-world use case

The line that we fit here is called the **ReLU activation function**, which we covered earlier. ReLu accepts input values and gives the maximum of 0 and the input value. This means that if the input values are positive, it returns as it is but if the input values in negative it returns 0. This is how we convert a linear function into a non-linear one.

With the current data we have, a perceptron is the simplest neural network we can implement. A perceptron consists of an input layer, a single hidden layer, and an output (*Figure 4.15*). It takes a vector of input values and performs a linear combination of values with the corresponding weight vector and the bias term. Then, the weighted sum is passed through an activation function, as discussed earlier:

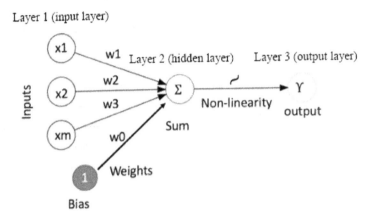

Figure 4.15 – Schematic representation of a perceptron

After receiving the data from the input layer – in this example, the GC content from multiple DNA sequences – the neurons process and pass it to the connected neurons through edges. Each edge has a weight associated with it. Formally, each neuron has inputs and each input is assigned a particular weight. The weighted sum of the input values and weights of the input nodes is passed through the activation function. A neuron may have a threshold (called bias) so that only if the aggregate signal crosses the threshold is the signal passed to the output.

Training a simple neural network (perceptron) starts with labeled data. In the case of the methylation value prediction example, the GC content of DNA samples 1, 2, 3, and so on, as shown in *Figure 4.13*, are input values, and their corresponding weights can either be initialized or determined from domain knowledge:

1. During training, we multiply each data point or GC content value with the initialized weights/ domain-determined weights + bias term.

2. Then, we pass the sum to an activation function, such as ReLu, to predict the methylation value for that data point.

3. Next, we multiply the value from *step 2* with weights of the edge that connects the hidden layer and the output along with the bias and use either ReLu or Sigmoid activation function to calculate the final output.

4. Then, we can calculate the loss to evaluate the predictions against the true labels.

5. During training, the derivative of the loss is calculated concerning the network weights using the chain rule.

6. The weights on the edges of the network are then optimized by updating them in the direction of the negative gradient using backpropagation, thereby minimizing the error between the predicted value and the output value.

7. This process repeats until we reduce the loss (sum of errors from all data points) to 0 or near 0. The lower the loss, the better the model.

8. Along with minimizing loss, in parallel, the network's accuracy is calculated using the validation dataset using cross-validation. This is mainly done to avoid overfitting the model.

9. As we saw earlier, multiple regularization techniques are available to reduce overfitting, such as dropout, Lasso, Ridge, and so on.

Similarly, we can build another perceptron by using another feature (for example, the k-mer frequency). It will look the same as it did previously except the features will be different. Furthermore, we can build a bigger neural network by combining several of these smaller neural networks, called **multi-layer perceptrons (MLPs)**, and adding more features to our methylation prediction model.

DNN architectures

Several types of neural network architectures are used for different use cases and data types. They can be broadly divided into the following categories:

- **Feed-forward neural networks (FNNs)** can stack up multiple perceptrons to build DNNs

- **Convolutional neural networks (CNNs)** are used for object detection, image classification tasks, and other computer vision tasks

- **Recurrent neural networks (RNNs)** are commonly used for sequential data and are mainly employed in NLP and speech recognition-related tasks.

- **Graph neural networks (GNNs)** are mainly used to leverage the structure and properties of the graph

- **Generative Adversarial Networks (GANs)** are commonly used architectures for learning complex patterns and also for generating synthetic data

- **Autoencoders** are mainly used for unsupervised learning

The following section briefly summarizes each of the different types of DNN architectures.

FNNs

So far, we have seen that, in a typical neural network, each layer generates some outputs that act as an input to the next layers, and so on. We call this an FNN, a standard neural network, or a vanilla neural network. This type of network best works on a structured dataset where well-defined features are available for each data point – for example, GC content, k-mer frequency, and so on. FNN has fully connected node layers, where each neuron in the first layer is fully connected to all neurons in the next layer and so on (*Figures 4.16*):

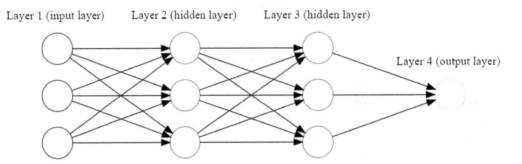

Figure 4.16 – An example of FNN with two hidden layers

The information passes from one layer to another in the forward direction in an FNN. There is no feedback from the later layers to the previous layers in this type of network architecture and the basic learning process of FNNs is the same as the perceptron. FNNs are widely used for generic prediction problems where it is hard to use classical ML algorithms.

CNNs

A CNN is a type of FNN that consists of one or more convolutional layers. They differentiate themselves from the other architectures by their application in image, speech, audio, and DNA sequence classification because they are good at finding patterns in the input images such as lines, curves, circles, and more complex patterns. We have a separate chapter on CNNs, so we will only briefly talk about the key components of CNNs here.

The core of the CNN architecture consists of three main layers: the **convolutional layer**, the **pooling layer**, and the **fully connected (FC) layer** (*Figure 4.17*):

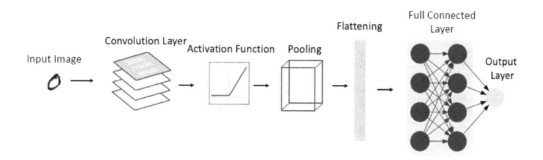

Figure 4.17 – Overview of the CNN architecture

The convolutional layer is the first layer after the input layer and is the core building block of a CNN. A convolutional layer is typically followed by one or more convolutional layers stacked on top of one another in a sequence (*Figure 4.17*). These convolutions allow CNNs to learn hierarchical features from the input data because each convolutional layer can recognize simpler features, such as shapes. The deeper the network goes, it can recognize even more complex features. They detect patterns in the input data with the help of filters, which are small matrices with a defined number of rows and columns. The filters scan the inputs and calculate the weighted sum of input values and filter weights and produce an output. You can imagine this as **positional weight matrix** (**PWM**) of a motif. The output of the convolutions is passed onto an activation function such as ReLU to bring non-linearity to the model. Next to the convolutional layer is the pooling layer, whose job is to reduce the dimensionality of the data by making the feature space smaller and the model more tolerant. Finally, the fully connected layer in the CNN flattens the feature matrix; the rest of the architecture is like a regular FNN.

Considerable attention has been paid to the application of CNNs in genomic problems, including sequence analysis, DNA sequence classification, gene probability, making predictions when binding DNA and RNA, and so on. One advantage of using CNNs over traditional ML methods is that you can automatically extract features and eliminate the need for manual feature extraction. In a typical CNN for a genomic application, such as in the example that we saw earlier (predicting methylation values), we can typically provide a set of DNA sequences, such as 400 bp, around the methylation sites and allow the CNN to extract the features using pre-defined filters. Then, we can use those features in the model, along with the methylation values for each of those DNA sequences. This can be followed by a series of pooling and fully connected layers to predict the methylation values. We will learn more about this in the next chapter.

RNNs

An RNN is a type of architecture that has a hidden state that captures the historical information until the current state (timestamp). A typical RNN has the input in the form of sequential data, a hidden internal state, a memory state that retains the historical information of the previous observations that gets updated every time it reads the next sequence of data in the input, and finally, the internal hidden state, which will be fed back into the model (*Figure 4.18*):

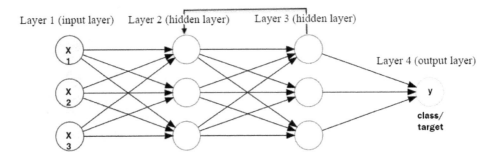

Figure 4.18 – RNN basic architecture

Let's say we want to understand the gene function only using the sequence. At the time of writing, conventional methods are available such as BLAST, functional annotation, phylogenetic analysis, and so on. This problem can be solved automatically using RNNs, which will take the input sequence and provide insightful information about that sequence – in this example, the function We have separate chapter dedicated for RNN and so we will skip the details in here.

GNNs

A GNN is a type of neural network architecture that is best suited for graph data. While typical neural networks work on array data as input, GNNs work with graphs. GNNs are one of the hottest topics in DL because of their huge popularity and their application in the many domains of life sciences. Graphs are everywhere; real-world objects are often defined in terms of their connections to other things. A set of objects, and their connections between them, are naturally expressed as a graph. GNNs work by transforming all attributes of the graph (nodes, edges, global context), as shown in *Figure 4.19*:

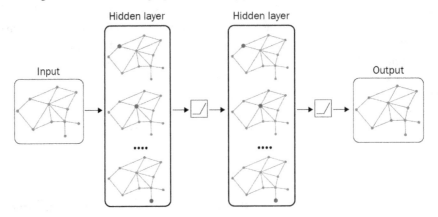

Figure 4.19 – Simplified GNN architecture

Recent advancements have increased GNNs capabilities and their expressive power and because of that, more and more fields are leveraging GNNs for solving complex problems. Some examples of GNNs include antibacterial discovery, physics simulations, fake news detection, recommendation systems, and so on. The simplest GNN is where we learn new embeddings for all graph attributes (nodes, edges, and global). This GNN uses a separate MLP on each component of a graph, and we call this a GNN layer.

GANs

A GAN is a new architecture type that uses two neural networks (generator and discriminator) that compete against each other (*Figure 4.20*):

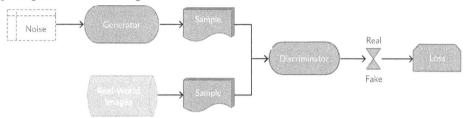

Figure 4.20 – GAN architecture

The generator's job in the preceding figure is to generate the synthetic realistic training data, while the discriminator determines if the authenticity of that generated data is fake or real by comparing it with the real training dataset. Because of their improved accuracy with classification tasks, GANs are used in many domains, including genomics.

Autoencoders

An autoencoder is a type of neural network architecture that is mainly used for unsupervised learning, which tries to copy the input data to an output in a non-linear fashion, unlike PCA, which is mainly linear. The two main components of an autoencoder are the encoder and the decoder (*Figure 4.21*):

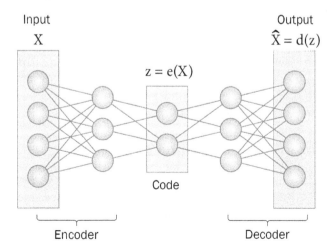

Figure 4.21 – Architecture of an autoencoder

> **Credit:**
> *bioicons.com* (`https://bioicons.com/`)

The encoder's job, as shown in *Figure 4.21*, is to compress the input to a low-dimensional code or latent-space representation, while the job of the decoder is to reconstruct the output from this representation. An autoencoder is designed to find low-dimensional space in the input data but reconstructs original information from the input in an uncompressed manner. Autoencoders have several applications, ranging from data denoising to dimensionality reduction and anomaly detection.

Now that you have some understanding of DLs, the neural network architecture, and the different neural network architectures available, let's see how these are all useful to the field of genomics – specifically, the DL workflow and the broad application of DNNs for genomics.

DNNs for genomics

The field of genomics has benefitted primarily due to the technological advancements in NGS, which can generate genomics data at a low cost and scale. This opened opportunities for many areas of science, such as bioinformatics and ML. Bioinformatics, the field of computational algorithms used to process biological data, was hugely successful in genomics, enabling rules to mine insights from the data. ML methodologies have a huge influence on genomics for solving some complex biological problems such as gene signatures, functional genomics, gene interactions, and so on. However, the current advances in genomics technologies, coupled with ever-growing genomics data, have created lots of new challenges that require highly accurate and sophisticated algorithms.

DL, a component of ML, has made significant strides in the areas of speech recognition, computer vision, machine translation, NLP, and so on. Because of DL's ability to solve complex tasks with high accuracy compared to traditional methods, DL is well suited for solving problems in genomics. The application of DL for genomics is relatively new, and this was made possible mainly because of technological advancements in sequencing technologies that can generate massive amounts of big genomics data, designing sophisticated DL algorithms, and the availability of next-generation hardware capabilities such as GPUs and TPUs. As such, several DL algorithms have been successfully applied to genomics data, such as disease detection, subtype classification, treatment classes, and so on. Compared to shallow ML methods, which rely on manually extracting features, DL models will do multiple transformations of features because of their unique architecture.

Deep learning workflow for genomics

The first step in applying DL to genomics is checking the availability of raw data, which can either be generated or extracted from multiple sources and preprocessed. As seen previously, the input of a DNN is real values and in the case of DNA sequences, the four nucleotides (A, T, C, G) can be one-hot encoded as [1,0,0,0], [0,1,0,0], [0,0,1,0], and [0,0,0,1]. The target labels for this data can either be human-annotated or experimental results. Similar to ML, the input data is split into training, validation, and testing datasets and are used for model training, model validation, and model evaluation, respectively. Again, the data split can depend on real-world scenarios. For example, you can keep most data for training (70%), a few data points for validating the model (10%), and 20% for testing (*Figure 4.22*):

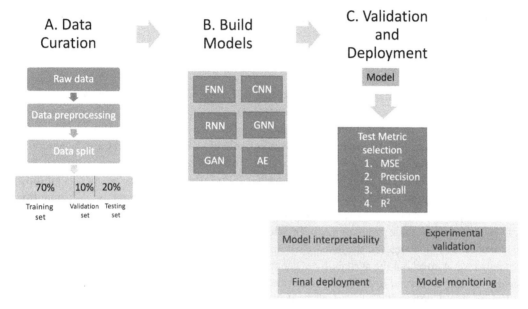

Figure 4.22 – DL workflow for genomics

After data splitting, the next step is model selection, which depends on the type of problem. Suppose you are interested in predicting the type of DNA sequence (enhancer or not) based on the domain knowledge that regular elements are spatial invariant, indicating that the CNN may be a good model for this kind of problem. Once you have decided on a model, you can use the training dataset to train the model and the validation data for model and hyperparameter selection based on a set of metrics, such as RMSE and others, as discussed earlier. We select the optimal model that has the lowest loss and use that to make predictions on the dataset with the help of evaluation metrics.

Once the model has been evaluated successfully on the test data, it can be used for two different purposes: understanding what biological mechanisms or features are learned by the model (model interpretability) and making predictions on unseen real-world data (model predictions) (*Figure 4.22*). Many times, genomic researchers are interested in features from the evaluated model instead of the model predictions. Model predictions are relevant if you are trying to build a model to make some predictions. In the case of life sciences, often, it is possible to validate the model predictions through wet-lab experiments. After validation, the model can then be deployed to make predictions on an unlabeled dataset to predict labels (*Figure 4.22*).

Broad application of DNNs in genomics

DL represents a state-of-the-art approach in many fields. Similarly, a growing number of genomic problems are currently using DL. Among the broadest applications of DNNs for genomics is functional genomics, which includes DNA sequence specificity predictions, predicting methylation values, gene expression predictions, controlling splicing, and so on. The following are some of the broad applications of DNNs in genomics.

Gene expression prediction

The completion of the **Human Genome Project** in 2003, along with the recent completion of the **Telomere to Telomere (T2T)** effort to sequence every nucleotide of the human genome by international research, allowed the research community to better understand the genes that control human health and development. Gene expression plays a key role in cellular machinery and is generally considered a proxy for the cellular state. For example, predicting gene expression accurately has far-reaching consequences in basic research as well as in industrial applications, such as guiding wet-lab experimentation to those settings that are likely to generate the best results, high-quantity recombinant protein expression, strain engineering, drug production, disease classification, and so on. Even though incredible progress has been achieved in this area, gene expression prediction is still one of the key challenges in the field of genomics. The idea of using DNNs for gene expression predictions is not novel but the availability of genomic data, coupled with recent advances in DNNs, provides an unprecedented opportunity to develop predictive models using gene expression data for disease classification. DNNs, with their ability to map the input-output data, are suited for gene expression-based disease classification. For example, CNNs were successfully used to classify Alzheimer's disease (`https://link.springer.com/article/10.1007/s00521-021-05799-w`), DNNs were used for skin cancer classification (`https://www.ncbi.nlm.nih.gov/pmc/articles/PMC8160886/`), and so on. When it comes to gene expression predictions, DNNs outperform other methods such as linear models, molecular thermodynamics, Bayesian networks, and so on because of their higher accuracy.

SNP prediction

Even though the human genome has 3 billion base pairs, only 0.1% genetic variation exists between individuals. The most common cause of this genetic variation is a change of a single base pair in the DNA sequence, which we refer to as **single nucleotide polymorphisms (SNPs)**. Many of these single base-pair changes have no impact on human health. However, some of these changes have important biological effects and contribute to some genetic diseases in humans. As such, SNPs are commonly used to detect disease-causing genes in humans, predict a person's response to drugs or their susceptibility to developing the disease, classifying complex diseases using Genomics SNP data (`https://www.sciencedirect.com/science/article/abs/pii/S1568494620306566`) and so on.

Protein structure predictions

One of the most recent success stories of DNNs in functional genomics is protein structure predictions. So, what exactly is the protein structure prediction? It involves modeling the relationship between the amino acids of a protein and its corresponding 3D structure. Deciphering the structure of the protein is widely considered one of the foundational problems of biochemistry and computational biology. DeepMind's AlphaFold sent shockwaves in the protein structure prediction competition (**Protein Structure Prediction Center (CASP)**) when it achieved an accuracy of >90 and took first place by a large margin. By 2020, AphaFold's performance is even more impressive, and it is now considered the go-to model for predicting protein structure.

Regulatory genomics

Regulatory genomics is the study of gene regulatory elements such as promoters, enhancers, silencers, insulators, and so on. They play an important role in gene regulation and hence functionally characterizing them is very important. In addition to these gene regulatory elements, identifying sequence motifs in DNA and RNA regulatory regions is key since they represent target sites of a particular regulatory protein, such as the **transcription factor** (**TF**). A variety of techniques are currently available to functionally characterize these gene regulatory elements and sequence motifs. Along with ML algorithms, DNN architectures such as CNNs and RNNs are successfully applied for regulatory genomics applications.

Gene regulatory networks

Gene regulatory networks (**GRNs**) are defined as networks that are inferred by gene expression data. GRNs are an exciting area of functional genomics and represent causal relationships between the regulators and the target genes. GRNs are important to understand the causal map of network interactions, molecular marker detection, hub gene detection, and so on. It is now routinely possible to perform high-throughput sequencing on any species of interest and generate gene expression data, though it is still a challenge to infer regulatory relationships between TFs, binding sites, and potential gene targets. DNNs, especially CNNs and GNNs, have proven to have much success in building GRNs (`https://link.springer.com/chapter/10.1007/978-3-030-05481-6_11`).

Single-cell RNA sequencing

Single-cell **RNA sequencing** (**RNA-seq**) is a relatively new technology but it has already revolutionized RNA-seq because of its incredible success and widespread applications, particularly in clinical diagnostics. This is possible because this technology can reveal the heterogeneity of tissues. Single-cell RNA-seq enables gene expression measurements in individual cells, thereby enabling cell-type clustering. Despite its huge success, biological inference remains the major limitation because of the sparse nature of the generated data. In addition, there is a large volume of dropout events in the data. DNNs – in particular, GNNs – have been successful in deconvoluting the node relationships in a graph through neighbor information propagation (`https://www.nature.com/articles/s41467-021-22197-x`).

Introducing deep learning algorithms and Python libraries

DL is an umbrella term that represents the different neural network architectures, along with the libraries for building the same. Unlike ML, the requirements for DL are quite diverse and the vast number of resources can be intimidating for those either looking to get into this field or those who have already been into it. Let's look at the top DL libraries and how they can be leveraged to build DL models. There are a few DL libraries that are currently available for building DL models, but we will only highlight the most popular and widely used libraries for building a variety of DL models and architectures – TensorFlow, PyTorch, and Keras. In addition, there are genome-specific DL libraries that are also available. It is beyond the scope of this chapter to go into details about each of these DL libraries, but we will provide fundamentals that are sufficient for building DL models on several use cases, as we will see in this chapter and the subsequent chapters.

General deep learning libraries

Let's start with the common DL libraries and understand how they differ from one another concerning building DL models and different neural network architectures.

TensorFlow

TensorFlow (https://www.tensorflow.org/) is the most popular open source DL and ML library written in several programming languages such as C++, Python, and CUDA. It was developed by the Google Brain team. It uses multi-dimensional arrays called tensors, which are well suited for large numerical computations. It provides scalable production with static graphs, which are great for inference purposes.

PyTorch

PyTorch (https://pytorch.org/) is becoming very popular recently because of its ease of use. It is an open source DL framework that accelerates the path from research prototyping to production deployment. It originated from an earlier project called Torch, an open source ML library that is no longer in active development. One thing to note is that even though Torch and PyTorch share some commonalities, PyTorch is its framework. The two key purposes of PyTorch are that it is a replacement for NumPy to use the power of GPUs and TPUs and it is an automated differentiation library that is useful for building DL models. The reason why GPU and TPU computing is important here is that a graphics card is excellent at doing basic calculations in parallel, which is key for the success of DL models. PyTorch acts as an automatic differentiation library, which is crucial for predicting and improving errors or losses when implementing ANNs.

TensorFlow and PyTorch are by far the two most popular libraries for building DL models today. Which library you choose can depend on several factors, such as style, model, and the project goal, as described briefly next.

Style

PyTorch is suited to those who are Python programmers as it is easy to learn and work with it. On the other hand, TensorFlow is suited to those who are hardcore programmers because it supports multiple languages such as C++, Java, Swift, and so on.

Model and data

Many of the pre-trained models, such as BERT and DeepDream, may be available in only one library or the other.

Model deployment

Where do you want to deploy your model? If you want to deploy a model on mobile devices, then TensorFlow is best because of TensorFlow Lite and its Swift API. In contrast, if you are interested in using the cloud for model deployment purposes and have a cloud preference, then TensorFlow Lite is tightly integrated with Google Cloud, whereas PyTorch is integrated well with TorchServe on the AWS cloud. When deploying models, you can either use TensorFlow Serving or TensorFlow Lite, depending on the application. We will discuss this in *chapter 11, Model Deployment and Monitoring* of this book.

Keras

Keras is a high-level user-friendly architecture API written in Python for building DL models. It simplifies the process of building DL models with its easy-to-use framework. Since it is built on top of TensorFlow, it can run on large clusters of GPUs or TPUs and supports large-scale computing.

Similar to how we compared TensorFlow and PyTorch, now, let's compare TensorFlow and Keras.

The TensorFlow library provides a low-level API for building and deploying the model. However, its learning curve is steep and not user-friendly. It is best suited for DL research, building complex DL models, working with huge datasets, and high-performance models. In contrast, Keras is user-friendly as it allows you to build models quickly and easily with less code in Python, allowing for quick prototyping. Furthermore, Keras is less error-prone and can build more accurate models compared to TensorFlow because of how it operates within the limits of its frameworks. For example, Keras sacrifices speed for accuracy and user-friendliness. However, it is not well suited for building complex DL models and errors are hard to debug in Keras. For our hands-on section in future chapters we will be using the Keras framework because of its simplicity and ease of use. For now, let's get familiar with Keras basics.

Here is the Keras tutorial that I mentioned to you in my previous message. Can you please make sure to have this inserted before "Deep learning libraries for genomics".

Deep learning libraries for genomics

So far, we have looked at the general DL libraries, but some specific genomic DL libraries are also available, as we will briefly describe in the following subsections.

DragoNN

DragoNN (`https://kundajelab.github.io/dragonn/`) is a library that is meant for teaching and learning about DL for genomics. The goal of DragoNN is to enable genomics researchers to get started with DL and for DL practitioners to apply the genomics of DL. The library provides a platform for generating data based on simulations, developing DL models, model interpretation, and so on. DragoNN has Jupyter Notebook support and a command-line interface for both building models and interpreting them on user-defined data.

Kipoi

Often, it is not necessary to build models from scratch. Instead, you can just use pre-trained models that already exist. Kipoi (`http://kipoi.org`) is one such resource that contains many ready-to-use pre-trained models for genomics. It consists of around 2,200 pre-trained models covering predictive tasks in functional genomics such as transcriptional and post-transcriptional gene regulation, and so on. It currently has a Python API and is accessible via the command line and R. In addition to using the pre-trained models from the Kipoi repository, you can retrain them with additional datasets or fine-tune the model to suit your needs.

So far, you have seen a variety of topics detailing DL for genomics. Now, let's summarize this chapter.

Summary

This chapter introduced you to DL, a subcategory of ML that leverages artificial neural networks to mimic human brains and perform automated tasks without human intervention. DL has certainly come to the fore in the last few years because of the incredible advancements in the availability of big data, sophisticated algorithms, and improvements in computational hardware such as CPUs and GPUs. We started this chapter by understanding why there is a need for sophisticated algorithms to mine insights from ever-growing genomics data and how DL, using DNNs, can fill that gap. The anatomy of the neural network architecture, along with the key components of neural networks, was introduced. Understanding these key concepts is important to be able to build a solid foundation for DL concepts, as well as understand how they relate to genomic applications. Then, you were introduced to the different neural network architectures, such as CNNs, RNNs, GANs, GNNs, and autoencoders, and understood how they are different from FNNs and their application to genomics. Finally, we looked at the most popular DL frameworks, such as TensorFlow, PyTorch, and Keras, and looked at some differences between them.

DL concepts were introduced at a detailed level in this chapter to give you a taste of what's to come in the upcoming chapters. In the next few chapters, we will look at each of the neural network architectures, such as CNNs, RNNs, Autoencoders, GANs, and so on, in detail. In the final section of this book, we will learn how to opertionalize DL models.

In the next chapter, we will start our DL journey with CNNs.

## 5

Introducing Convolutional Neural Networks for Genomics

In recent years, **deep learning** (**DL**) has emerged as a prominent technology in solving complex problems in various domains. Among DL algorithms, **convolutional neural networks** (**CNNs**) dominate the current DL applications because of their incredible accuracy in **computer vision** (**CV**) and **natural language processing** (**NLP**) tasks. A CNN is a type of **neural network** (**NN**) architecture that is used for unstructured data and was originally designed to fully automate the classification of handcrafted characters. Some popular applications of CNNs include facial recognition, object detection, self-driving cars, auto-translation, handwritten character recognition, X-ray image analysis, cancer detection, biometric authentication, and so on. Compared to **feed-forward NNs** (**FNNs**), which we learned about in the previous chapter, CNNs process multiple arrays using convolutions within a local field, like perceiving images by eye. Thanks to next-generation hardware such as **Graphics Processing Units** (**GPUs**) and **Tensor Processing Units** (**TPUs**), the current CNN models can be built rapidly and cheaply. In addition, the introduction of DL libraries and frameworks such as TensorFlow, Keras, and PyTorch has enabled the models to be built and learned easily.

In genomics, **machine learning** (**ML**) methods have been widely used for solving some complex challenges, as seen in *Part 1* of this book. With the availability of massive amounts of data generated from high-throughput sequencing technologies such as **next-generation sequencing** (**NGS**), it was evident that it is beyond the reach of state-of-the-art technologies such as bioinformatics and ML to process this data automatically and provide knowledge on prediction-based analysis and biological insights. To address this, various CNN architectures were developed to process non-image data such as DNA sequence data for various applications such as cancer detection, predicting phenotypes from genotypes, gene expression prediction, identifying DNA- and RNA-binding motifs, predicting enhancers, and so on. The goal of this chapter is to introduce you to CNN architectures and applications as they relate to genomics. By the end of this chapter, you will have a good understanding of what CNNs are, different CNN architectures, why they are important in DL, and what some of the popular applications of CNNs in genomics are.

As such, here are the topics that will be covered in this chapter:

- Introduction to CNNs
- CNNs for genomics
- Applications of CNNs in genomics

Introduction to CNNs

Just to refresh your memory, FNNs are fully interconnected NNs where all nodes in the preceding layer are connected to every other node neuron in the next subsequent layer, and so on (*Figure 5.1*). Each edge or connection has a weight, that is either initialized randomly or derived from domain knowledge and ultimately learned by the algorithm during model training. The weights are then multiplied by the input values from all the node's neurons, and then the sum of all the nodes in the input layers is then passed on to the next layer, along with a bias that is then used by an activation function to signal whether that output will be passed on to the next layer or not. The process repeats in each layer until the final output layer, which has one to many neurons, depending on the type of learning and whether generating a prediction or classification. FNNs work great for structured data where you have M features and N samples as input:

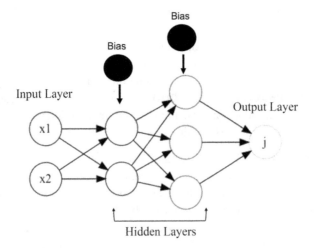

Figure 5.1 – Typical architecture of an FNN

However, using FNNs for unstructured data such as images, audio, text, sequence, and so on would be very challenging. Let's understand this through an example.

Say our input is a 32-pixel by 32-pixel grayscale image where every pixel in the grayscale image is represented by a value between 0 to 255. Here, 0 indicates black, 255 represents indicates white, and any values between 0 and 255 indicate various shades of gray. Since the grayscale image has only one channel, the image can be represented as *32 x 32 x 1 = 1024*, and consequently has 1,024 nodes in the input layer of the FNN. Suppose our next layer (hidden layer) has 100 nodes, and since this should be fully connected to the previous layer, we will have 1,024 x 100 = 102,400 weights for the first 2 layers (input and the first hidden layer). Since we know by now that a complex problem such as this requires multiple hidden layers in the FNN to map the inputs and the outputs in the training data to generate an accurate model, as such, we now have a problem with too many parameters in the FNN, which makes the training process very complex because of the increased dimension space. In addition, it makes the learning process slower, uses more resources, and increases the chances of overfitting.

If we are to use the color images, the problem is further compounded because the color images have 3 channels (red, green, and blue) where each color is represented by a channel, and so in total there are 3 values and 32 x 32 x 3 = 3,072 values. Correspondingly, the number of weights for such an image is 3,072 x 100 = 3,072,000. FNNs cannot scale to handle these kinds of images, so there needs to be a better scalable architecture. Another limitation of FNNs with image data is that the 2D image is converted to a 1D flattened vector, and so the spatial relationship of the different pixels is completely lost. So, there needs to be a better NN architecture that can keep spatial relationships and process unstructured data.

What are CNNs?

CNNs are a special type of NN architecture that is widely applied for unstructured data such as images, audio, video, text, and so on. CNNs are great networks for analyzing data with spatial dependencies such as images, audio, DNA sequences, and so on. They work well on DNA sequence data. The main applications of CNNs include image classification, NLP, signal processing, and speech recognition. CNNs have a series of convolutional layers, which allows them to automatically extract hierarchical patterns in the data. CNNs are currently being used in genomics tasks where local patterns are very important to the outcome—for example, the detection of conserved motifs to identify blocks of genes in a DNA sequence or binding sites of a protein such as a **transcription factor** (**TF**). Let's dive deep into the wonderful world of CNNs in the following sections.

Birth of CNNs

The first idea of a CNN came from Hubel and Wiesel when they reported in 1959 a marked change in neuron activity when a bright light was passed at a particular angle and a particular location to a cat (`https://www.ncbi.nlm.nih.gov/pmc/articles/PMC1363130/`). It was shown that complex neurons receive inputs from multiple simple neurons in a hierarchical manner. For these findings, they were awarded the Nobel prize in 1981. This report inspired Fukushima, who then developed the first CNN architecture termed *Neocognitron* to recognize handwritten Japanese characters (`https://www.rctn.org/bruno/public/papers/Fukushima1980.pdf`). CNNs have a long history, which primarily started with LeCun, who is widely credited for his contribution to advancing CNNs (`https://www.persee.fr/doc/intel_0769-4113_1987_num_2_1_1804`). However, the breakthrough for CNNs came when a CNN did extraordinarily well against other models at the **ImageNet Large Scale Visual Recognition Challenge** (**ILSVRC**) competition.

So, what is a CNN, how does it work, and why is it powerful for unstructured data? Let's start with understanding CNN architecture.

CNN architecture

Unlike an FNN, the architecture of a CNN is slightly complicated. It consists of an input layer for feeding in the inputs, a convolution layer for performing convolutions, a pooling layer for downsampling the data, a fully connected layer (which is similar to an FNN), and finally, an output layer for making predictions (*Figure 5.2*). Let's discuss each of these in detail in the following section:

Input layer

As with any NN, the input layer represents the first layer of a CNN. A typical example of an input layer can be a grayscale image, or, in the case of genomic data, it is the sequence of DNA from a patient sample or an experimental sample. Unlike vanilla NNs, there is no need to flatten the input to a one-dimensional vector. Instead, we can provide the images or a DNA sequence directly to the network, which will help to preserve the spatial relationships better.

Convolutional layer

This is the most important layer of a CNN because this is what distinguishes a CNN from an FNN and other architectures. The main feature of CNN is it's use of convolution operations which replace the matrix multiplications in a typical FNN and it's ability to capture spatial information in the data (*Figure 5.2*). Let's understand how a convolutional layer works in a typical CNN:

- A convolutional layer consists of multiple filters (also called **kernels**) that are nothing but the matrix of weights—for example, a filter of size 9 (3 by 3) that is initialized with values randomly between 0 to 10.

- Next, convolution is done by placing this filter at the top-left corner in the case of the image and taking the dot product of the filter values and the input values of the pixel (remember—pixel values range from 0 to 255, but they are normalized to keep the values between 0 to 1) to calculate a feature map, and we will take a step size (stride) of $n=1$ to the right.

- The process is repeated until we reach the top-right corner of the image, and then we start over again from one down cell from the left of the image until we finish scanning the whole image.

- While convolving the filter, we calculate the dot product of the weights on the filter with the input values. For example, as shown in *Figure 5.2*, the input size of 5 by 5 (25 in total) is being scanned by a filter of size 9 (3 by 3), and the dot product of the filter values and the input values of the pixel is being calculated in the output array:

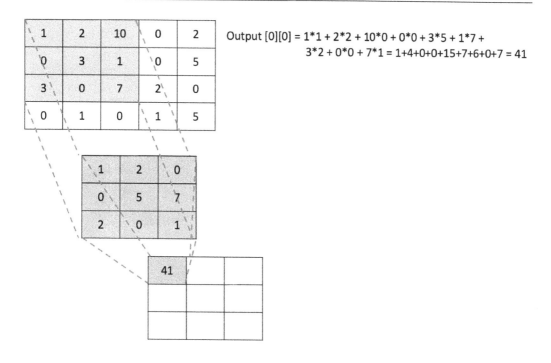

Output [0][0] = 1*1 + 2*2 + 10*0 + 0*0 + 3*5 + 1*7 +
3*2 + 0*0 + 7*1 = 1+4+0+0+15+7+6+0+7 = 41

Figure 5.2 – Illustration of a convolutional filter along with the input image

In the case of a DNA sequence as an input, after creating a one-hot encoding (*Figure 5.3*), the convolutions are performed similar to image data. Each filter chosen will act as a sliding window.

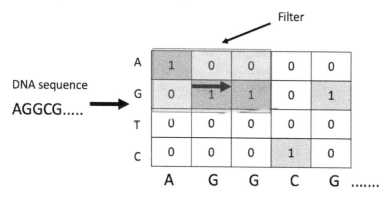

Figure 5.3 – Illustration of a convolutional filter along with the input image

Generally, multiple convolutional layers are present in a typical CNN architecture. The first few convolutional layers extract general features from their input data, and the subsequent layers extract more complex features. A non linear activation function such as ReLu or LeakyReLu will be applied to the linear combination of weights in the filter and input values.

Pooling layer

The next layer after the convolution layer is the pooling layer, which reduces the number of parameters, memory usage, and computational cost, and prevents overfitting in the network progressively (*Figure 5.4*). It is quite normal to have pooling layers between multiple convolutional layers. The pooling layer operates as a downsampling layer, and while doing so, computes either an average or a maximum value for a specified filter size and stride (slide is nothing but a sliding window size):

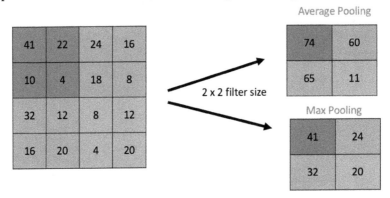

Figure 5.4 – Pooling layer schematic

For example, as shown in *Figure 5.4*, the max pooling takes the maximum value (74) from the window, which are 41, 24, 32, and 20, and the average pooling takes the average of all the values from the window, which are 74, 60, 65, and 11. There are no parameters need to be learned from the pooling layer.

Fully connected layer

The last layer before the output layer is the fully connected layer, and this generally lies between the pooling layer and the output layer. Please note that the output from the pooling layer gets flattened before feeding it to the full connected layer. All the nodes from the pooling layer are connected to the fully connected layer to generate the probability of predictions or classifications (*Figure 5.5*):

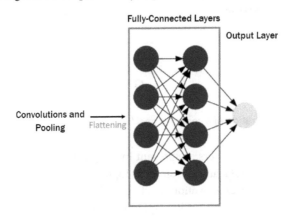

Figure 5.5 – Fully connected layer connecting to the output layer

The fully connected layer is the key since it takes the flattened values from the preceding convolutions and pooling layers, then behaves as a mini FNN, and then connects it to the output layer (*Figure 5.5*).

Output layer

This is like any other output layer as we have seen in a typical FNN. The output layer uses a function such as Softmax or Sigmoid to predict or classify the output depending on the problem.

So far, you have been looking at individual components of a CNN one at a time. Let's now visualize a full CNN end-to-end (*Figure 5.6*). As indicated previously, a typical CNN usually consists of multiple convolutional and pooling layers. The convolutional layers along with the pooling layers are very important as they are responsible for feature extraction, which is a powerful feature of a CNN, unlike other architectures. In this example, the input such as an image (here, the image representing 0) or a DNA sequence in the form of a one-hot encoding matrix is fed into the CNN model, which goes through several convolutional, pooling, flattening, fully connected and finally, gets predicted in an output layer:

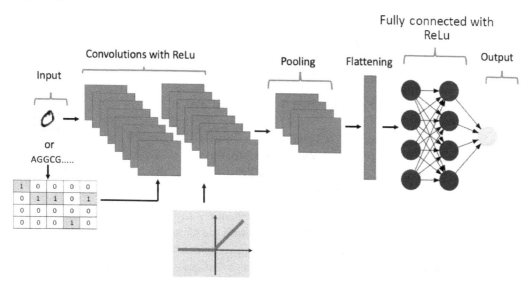

Figure 5.6 – A typical CNN with convolution layers

Now that we have understood what a typical CNN architecture looks like, let's take a small detour and learn about **transfer learning (TL)**, which is a powerful concept in CNNs and other NNs.

Transfer Learning

Transfer learning (TL) is where we leverage a pre-built model on a new task. In other words, a model that was trained on a particular task is reused on a different but related task to repurpose the existing model without building a completely new model from scratch. In TL, the learned feature from the pre-trained model is transferred to facilitate the prediction or classification of untrained datasets. It's like taking a model and using it for prediction purposes without training the model from scratch (*Figure 5.7*). Through the process of TL, we can achieve significantly higher accuracy compared to training a model from scratch with only a limited amount of trained data and labels:

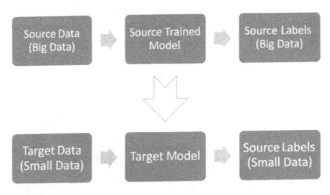

Figure 5.7 – How TL helps in CNN tasks

As shown in the preceding diagram, a CNN model is typically trained on datasets that are large so that the model can learn the features effectively compared to smaller datasets. Once this model is trained on this big data, then it can be used to retrain with smaller data specific to a particular problem, and then the model can not only be trained effectively but also produce highly accurate results.

TL is quite common in DL, especially for applications such as CV and NLP-related tasks where it is not always possible to get a large amount of data along with the target labels for model training. It is rare for researchers and scientists to build models from scratch; instead, they prefer to start from a pre-built model that has already been built for doing a similar task—for example, model that can classify different sequences and has learned general patterns of the data. The same can be said for genomics. Many times, it is not possible to build a model for genomics problems due to the absence of labeled data, smaller-sized samples compared to the number of genes, and so on. In this instance, the TL method can aid the genomics field by incorporating the complex features learned by the model on the trained data and using that to predict or classify the novel data. A few examples include predicting cancer types, identifying potential biomarkers for small or large untrained datasets, predicting gene expression, and so on.

CNNs for genomics

Even though CNNs are primarily used for unstructured data such as images, text, audio, and so on, they are also powerful tools for non-image data such as DNA. Unfortunately, the raw DNA sequence data cannot be provided to CNNs as input for feature extraction. It has to be converted to numerical representation before it can be used by CNN. The first thing to note for non-numeric data such as a DNA sequence is that you will have to first convert the 1D DNA sequence data to a one-hot encoded structure (*Figure 5.8*):

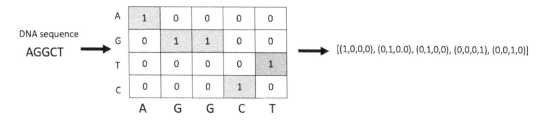

Figure 5.8 – Example of one-hot encoding for a DNA sequence

As shown in the preceding diagram, each nucleotide in the DNA sequences is represented as a one-hot vector: A = [1000], C = [0100], G = [0001], and T = [0010]. The one-hot encoded matrix can then be fed into the model for training purposes. Please note that one-hot encoding is not the only way of representing DNA sequences to a CNN. There is also label encoding in which each nucleotide (A,G,C,T) is represented by a unique index thereby preserving the positional information. For example A=1, G=2, C=3, and T=4. Let's try to understand why a CNN architecture is great for genomics problems using cancer prediction from **Single nucleotide polymorphisms** (**SNP**) variants from multiple patient samples as an example.

In the case of using FNN for cancer prediction, the first layer receives the SNP variants as input, and the weighted sum of each input value plus the "bias" (constant) is then passed through a non-linear function such as ReLu. This process repeats in the subsequent hidden layers until it reaches the output layer where it is transformed via sigmoid or softmax activation function to produce the hidden neurons output value. Although FNNs are powerful for this example, they are certainly not suitable for spatial or temporal datasets, and in genomics, there are several problems that are either spatial or temporal, and predicting cancer from SNPs is one of them. CNNs can address these issues. In the case of cancer prediction from the SNP data problem, SNPs are distributed according to a particular space pattern. So, a convolutional operation with predefined width and strides is performed on the input data. In this example (*Figure 5.9*), the convolutional operation is done using a kernel size of 1 by 3 and with a stride of 1. After multiple convolutions in one or more convolutional layers, the next layer is the max pooling layer, which takes the maximum of all values for each of those positions from kernel outputs using a predefined filter size. Pooling layers are then flattened and connected to a non-linear fully connected layer and finally to an output layer that has a binary class (cancer or no cancer) (*Figure 5.9*):

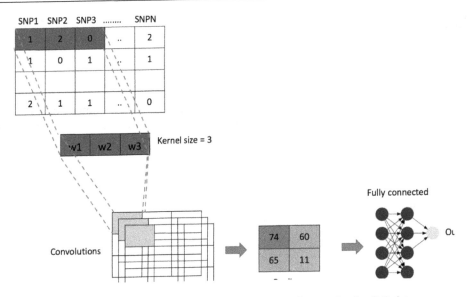

Figure 5.9 – Simple schematic of a 1D convolutional operation for SNP data

As you can see in the preceding diagram, CNNs are thus very powerful for spatial interactions and are translation-invariant, and so are routinely used in genomic applications where both spatial and temporal interactions are quite common. In the next section, we will learn some of the most popular applications of CNNs in genomics.

> **Note:**
> We will see an use case of of how to build CNNs for predicting the binding site location of a Transcription Factor (JUND) in *Chapter 9, Building and Tuning Deep Learning Models*. For now, let's look at the different applications of CNNs in genomics in the next section.

Applications of CNNs in genomics

Now that you understand how CNNs work in genomics with a simple example, let's look at some of their applications that are popular in genomics.

DeepBind

TFs are DNA- and RNA binding proteins that play a crucial role in gene regulation. Knowing the binding sites of these DNA- and RNA-binding proteins would help us to develop models and can help identify disease-causing variants. One way to infer the sequence specificities of these proteins is through **position weight matrices (PWMs)**, which can be used to scan the entire genome to identify potential binding sites of these DNA- and RNA-binding proteins. In addition, DNA- and RNA-binding protein specificities are measured by several high-throughput assays such as PBM, SELEX, and the most popular

ChIP- and CLIP-seq techniques. Some of the challenges associated with this data are that the raw data comes in several different quantitative forms, the data is often extremely huge, and each data generation methodology suffers from artifacts, biases, and limitations that hinder the identification of binding sites. DL techniques show more promise in capturing sequence specificities of these binding proteins. DeepBind (`https://pubmed.ncbi.nlm.nih.gov/26213851/`) is a CNN architecture that was built to predict DNA- and RNA-binding protein binding sites in the genome. It takes in noisy experimental data and outputs a binding affinity of a DNA- or RNA-binding protein to indicate how likely it is that the TF will bind to the sequence.

DeepBind takes in a set of raw sequences with lengths ranging from 14 to 101 base pairs, along with an experimentally determined binding value as a label or target. From this data, it calculates the binding score in four stages—the convolutional stage, which scans the sequences with a set of motif detectors (4xm matrices) such as PWMs. The next is the rectification stage, which isolates any patterns by shifting responses of the motif detector and clamping any negative values to 0 using the ReLu activation function (remember—ReLu takes in a value and gives either 0 or the maximum of the value, whichever is the greatest). Following the rectification is the pooling stage, where the model calculates both the average and maximum value for each motif detector. The max pooling will help the model get the longer motifs or patterns, whereas the average pooling will help the model pick up the combined effect of shorter motifs.

The pooled values are then flattened and fed into the fully connected layer (non-linear NN) to produce the scores. The rest of the steps are typical of any FNN, which is to use backpropagation, cross-validation, and so on to optimize the weights and improve the model accuracy. The optimized DeepBind model is evaluated on test data to predict binding affinities and model performance using metrics such as AUC. Using DeepBind, you can now simply change the nucleotide position of a protein of interest and see how that affects the binding affinity of a TF in terms of positive or negative binding. Since DeepBind can find relationships between mutations and the dishes that cannot be inferred from traditional methods, researchers have used DeepBind to see where certain mutations in the cholesterol gene can disrupt the binding affinity of a TF.

DeepInsight

DeepInsight (`https://www.nature.com/articles/s41598-019-47765-6`) is a CNN model that was built to extract features from genomics data. The idea of DeepInsight is simple. It takes non-image data such as genomics, transforms it into image data, and then feeds it into a CNN for classification or prediction depending on the problem. Instead of feature extraction and selection for collected samples (M samples x d features), DeepInsight finds a way of arranging closely related features into neighboring regions (d features x M samples), and all dissimilar features are kept further apart with the goal of learning complex relationships and interactions. Arranging similar elements together is a powerful method for uncovering hidden mechanisms such as pathways in biological systems. This approach of clustering similar features is more powerful than handling each feature separately, which ignores the neighborhood information that is key for spatial and temporal datasets. With this simple idea, it can transform any non-image data into feature map images, which can be a friendly representation of samples to a CNN architecture.

DeepChrome

Among several factors that control transcriptional regulation, histone modification is the primary factor. Histone nucleosome proteins and histone modification through methylation is a very common gene regulatory mechanism. Understanding of the combinatorial effects of histone modification and gene regulation can help in developing epigenetic drugs for cancer. Researchers are interested in inferring the gene expression from methylated data, and so multiple computational models, both rule-based, which try to capture the relationships between histone modification and gene expression, and ML models such as linear regression, support vector machines (SVM), and Random Forest, have been proposed for gene expression predictions based on histone modification. However, neither of these methods have been successful. DeepChrome (`https://academic.oup.com/bioinformatics/article/32/17/i639/2450757`) is a CNN-based architecture that was designed to capture the interactions among histone modifications and use that to predict gene expression. If you recall, we have seen an example of methylation prediction using manually extracted features in *Chapter 4, Deep Learning for Genomics*. DeepChrome, allows automatic extraction off complex interactions or most important features without us manually extracting the same as we have done before.

DeepVariant

NGS technologies have allowed scientists and clinicians to produce sequencing data rapidly, cheaply, and at scale and apply it broadly to fields such as health, agriculture, and ecology. Accurate detection of genetic variants (SNPs and indels) from sequencing data is important for scientists and clinicians for disease detection, identifying genetic disorders, and making discoveries. DNA sequencing of genomes to identify genetic variants involves two key steps:

1. Sequencing of the samples, which generates relatively short pieces of DNA (reads)
2. A variant caller that maps the reads against the known reference to identify where and how a sample's genome sequence is different from a reference genome

The current variant callers often struggle to accurately identify the correct variants and, in the process, generate lots of false-positive and false-negative variants. Accuracy of the variant detection is important because a false-negative variant indicates missing the causal variant for a disorder while the opposite (a false positive) means identifying the wrong variant. DeepVariant (`https://www.nature.com/articles/nbt.4235`) is a CNN-based framework that converts variant identification problems into image classification problems. It takes the input data, which is aligned sequences (reads) to the reference, and constructs multichannel images representing some aspect of the sequence—for example, one channel can be *read base*, the second channel is the mapping quality, and the third channel is *read supports allele*, and so on. Finally, DeepVariant generates one of three genotype likelihoods (0, 1, or 2) of given alternate alleles in each sample. Using this unique design, DeepVariant outperformed state-of-the-art variant detection tools. The model can be used across other genomes allowing nonhuman sequencing projects to benefit other genome projects. In addition, DeepVariant can be leveraged for other sequencing technologies and experimental designs highlighting the importance of this architecture.

CNNC

A **CNN for coexpression** (**CNNC**) (https://www.biorxiv.org/content/10.1101/365007v2) is a supervised CNN architecture that uses gene expression data to understand relationships between genes. Each pair of genes is represented as a histogram, and a CNN model is trained with training examples that have both positive and negative samples. Some examples of these include known targets of a TF, known pathways for a specific biological process, known disease genes, and so on. First, the gene expression levels from each pair of genes are transformed into 32 x 32 **normalized empirical probability distribution function** (**NEPDF**) matrices. Next, each NEPDF matrix is fed into the network as an input. The other input types can be DNase-Seq, PWM data, and so on. After going through the convolutions, finally, the output of a CNNC can either be a binary classification or a multi-class classification.

Summary

DL has made massive strides in several domains of life sciences and biotechnology, including genomics. CNN architecture is mainly designed for unstructured data. It accepts the input image or a DNA sequence (matrix of size m x n) as an input, extracts the features from the image, and does the prediction or classification through a series of hidden layers such as a convolutional layer, a pooling layer, a non-linear fully connected layer, and an output layer. CNNs do not require any separate feature extraction step and automatically derive features from the input data. CNNs have revolutionized the field of genomics because of their incredible accuracy and ability to process unstructured data, which is quite common in genomics.

In this chapter, we have looked at the history of CNNs, what they are, and the different components of CNN architecture. Later in the chapter, we understood how CNNs are being leveraged in genomics for studying complex problems such as gene expression, gene regulation, genetic variant detection, and so on. In the final section of the chapter, we have seen different CNN architectures for solving genomic problems. Hopefully, this chapter gave you a taste of CNNs and what they can do to solve some of the complex challenges in genomics. In the next chapter, we will learn about another incredible architecture that is even becoming more popular than CNNs for genomics: **recurrent NNs (RNNs)**.

6

Recurrent Neural Networks in Genomics

Deep learning (DL) models are so versatile that they can adapt to any input data distribution and, at the same time, generalize very well to previously unseen data. A variety of **deep neural network** (**DNN**) architectures have been designed to suit a particular task. For example, we saw how **feedforward neural networks** (**FNNs**) are good at making predictions from structured data, such as tabular data, in *Chapter 4*, *Deep Learning for Genomics*. We also saw how **convolutional neural networks** (**CNNs**) are good at making predictions from unstructured data such as images, audio, text, and DNA sequence data; we saw this in *Chapter 5*, *Introducing Convolutional Neural Networks for Genomics*. But what about sequential data? If you look around, we are currently flooded with a lot of sequential data. Some examples include financial data and DNA sequences. The most important type of sequential data is the time series data, which is a series of data points listed in time order. This data is key for applications such as speech recognition, sentiment analysis, language translation, and so on.

A **recurrent neural network** (**RNN**) is a DNN architecture that has a feedback loop to previous timestamps to make predictions of the future state of the input data. RNNs have revolutionized predictive analytics for temporal and sequential data. RNNs are now routinely used to address **machine learning** (**ML**) problems in domains such as **natural language processing** (**NLP**) for everyday applications such as language generation, named entity recognition, sentiment analysis, image captioning, text summarization, and more. The field of genomics, which consists of the most natural language ever – a sequence of nucleotides (A, G, C, and T) – is very well suited for RNNs applications, such as for predicting proteins from DNA sequences, predicting the binding domains of proteins, predicting the interaction between enhancers and promoters, predicting structural motifs, predicting base calls from sequencing instruments, optimizing coding sequences for increased protein production, predicting function, and so on. In this chapter, you will learn what RNNs are, how they are different from FNNs and CNNs, and how they are better suited for sequential data. By the end of this chapter, you will understand what RNNs are and why they are important in DL, the different types of RNN architectures and when to use what, and the different RNN applications in genomics.

As such, here is the outline for this chapter:

- What are RNNs?

- Introducing RNNs

- Different RNN architectures

- Applications and use cases of RNNs in genomics

What are RNNs?

Before we understand RNNs, let's refresh our memory and revisit how FNNs and CNNs work. In a typical FNN, you have an input layer, multiple hidden layers, and an output layer. After all the data is fed into the input layer, the information passes to the hidden layer. Then, the dot product of the input value and weight of each node is summed up, along with the bias term, which is turned into an activation function at each of the three nodes (*Figure 6.1*). The activation function can be binary, sigmoid, ReLu, LeakyReLu, or something else, as you learned in *Chapter 4, Deep Learning for Genomics*. Depending on the type of activation function, the value of the single node in the hidden layer is outputted:

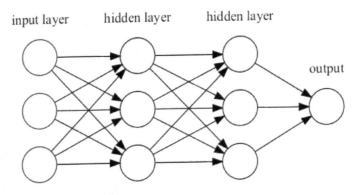

Figure 6.1 – A multi-dimensional input type FNN

The number of nodes in the output layer depends on the problem and the required output. For example, if you are trying to classify a DNA sequence based on mutations in each of the 10 different cancer types, the output layer will consist of 10 nodes, one for each of the relevant cancer types and the output will be a probability of each cancer type. In the case of supervised learning, we estimate how far we are with our predictions by using the *loss* function; then, the learning algorithm adjusts the parameters iteratively so that we have a highly accurate model with low or zero loss. As you learned previously, techniques such as backpropagation, gradient descent, batch normalization, regularization, and others are used to achieve this.

CNNs, on the other hand, can handle unstructured data such as images, audio, video, and DNA sequences and automatically extract patterns from unstructured data. CNNs take a sequence of DNA (in this example, a one-hot encoded matrix) as input and involve a series of convolutions that use several filters of predefined weights, then perform non-linearity using activation functions such as ReLU. Then, we perform either a max or average pooling on the activation matrix to reduce the dimensions of the data and extract the higher-level features from the input data. The output from the pooling layer is flattened and then connected to a fully connected layer that has a vector of fixed size.

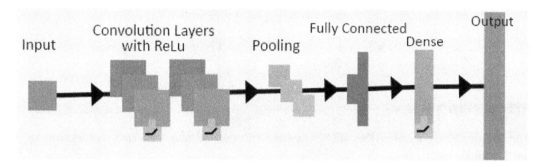

Figure 6.2 – Simple overview of a CNN

As shown in the preceding diagram, the CNN architecture consists of two convolutional layers with activation functions such as ReLu, a pooling layer, fully connected layers and finally an output layer.

Even though both FNN and CNN models have achieved better performance than conventional methods, they only focus on the current state without referring to the previous state to adjust their predictions. They both assume that each input is independent of the other and they take one input at a time and generate one output at a time.

However, everything in life is determined by a sequence of events that are both sequential and temporal. The data, such as audio, video, or text, has a sequential order of events. For example, video is composed of a series of video frames, audio is composed of a series of sound elements, DNA consists of a series of nucleotides (A, G, C, and T), and so on. In sequential data, each input is dependent on the previous input. For example, to map the audio content to a text transcript, to predict the sentiment (positive, negative, or neutral) of the message from the input text, or to predict proteins from the exact location on a DNA sequence at a given state, we must have information about the previous event. So, let's understand why FNN and CNN cannot be leveraged for this kind of data behavior:

- An FNN can be used on this type of data and can feed the entire sequence, but the size of the input and output would have to be fixed, which is not ideal for this type of data. What if the most important events of the sequence are outside of that window?

- Can we use CNNs for this type of data? CNNs are good for capturing spatial information of the data. They use convolutions, which rely on a fixed-size input; for example, convolutions on a specific number of characters in an input DNA sequence. This will be a problem in cases where you're unsure about the size of the subsequence. This problem can be addressed with a different-sized convolutions to suit the different-sized sequence data. However, this will lead to more complexity in the network as it will lead to more fully-connected layers. CNNs can also address nonlocal relationships in the data since they take adjacent pixels or sequences into account. However, CNN models only capture the current state and not how the previous state will influence the current state for future predictions.

What we need is a network that can not only identify nonlocal relationships but can also remember and carry this information over a long distance.

Introducing RNNs

To address the limitations of FNNs and CNNs regarding sequential data, we need a network that can meet 2 requirements.

1. It can takes sequences of non-fixed lengths, one element of the sequence at a time.

2. It must not only identify the nonlocal relationships in the sequence but also remember the most important events that happened before.

This idea led to the development of RNNs, which are a variant of DNN with a feedback loop (hidden state) that can feed the results back into the network and make them part of the final output (*Figure 6.3*):

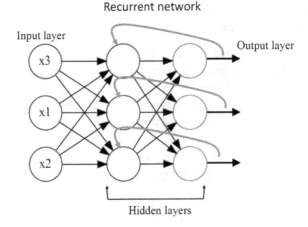

Figure 6.3 – Recurrent neural network

RNNs capture previous observations or historical events up to the current timestamp and because the hidden state of the current stamp is the same as the previous timestamp, the computation is recurrent (hence why they are referred to as RNNs):

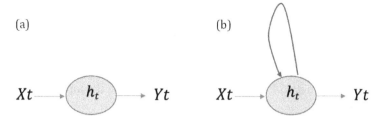

Figure 6.4 – Difference between a standard neural network (a) and a recurrent neural network (b)

As shown in *Figure 6.4 a*, in standard neural network (FNN), the network takes inputs, X_t, and passes them to the hidden layer, h_t, and then generates the final output, Y_t. In an RNN (*Figure 6.4 b*), the input vector, X_t, is fed into the network one element at a time but instead of outputting, it creates a feedback connection from the hidden layer, h_t, to itself so that it gets access to previous information at all the timestamps. Optionally, the network can also generate the output as shown in *Figure 6.4b* Y_t. But how do they feed the results back into the network and how do they store information from a previous observation in memory? To understand this, let's look at the RNN architecture in detail now.

Another good way of illustrating how RNNs work is to explain it with an example: Imagine you have a standard FNN and give it a DNA sequence (ATGCGAG) and it processes one nucleotide at a time but by the time it reaches the last nucleotide (in this example 'G') it has forgotten everything about other nucleotides 'A', 'T', 'G', 'C', 'G', 'A' and FNN can't predict what nucleotide would come next. This information is important for sequential data such as DNA sequences because there is a structure to the sequence

An RNN, however, can remember those nucleotides because of its memory state (which you will understand in the next section). It produces the output and is copied back into the network using a feedback loop.

Now that you have some basic understanding of what RNNs are, let's look at the RNN architecture.

How do RNNs work?

A standard RNN has an input layer, a feedback loop, and an output layer (*Figure 6.5*):

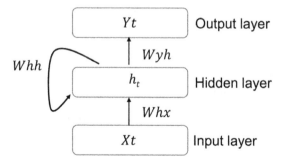

Figure 6.5 – A schematic diagram of a simple RNN

Let's look at simple RNN in more detail:

- The RNN accepts the input vector into the network and passes it through a hidden layer (RNN block), which then stores that information in memory and gives an output vector. An RNN's main feature is that the result in the output vector is not only dependent on the input vector that gets fed into the network via the timestamp but also on the input vectors from all the previous timestamps.

- A typical RNN class has some hidden state (which is nothing but the output from previous timestamp that is looped back into the network) that gets updated at each timestamp until the input sequence is exhausted using a feedback loop in the RNN cell, which gets updated in every iteration (*Figure 6.5*).

So, what is the functional form of this recurrent relationship that we are computing in the RNN block?

The recurrence relationship function, which is the key part of the RNN block, is shown in *Figure 6.6*:

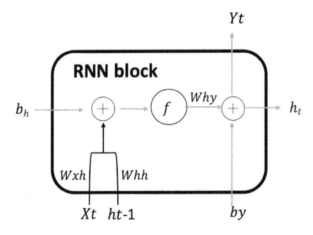

Figure 6.6 – Architecture of a typical RNN

While processing the input vector, the network calculates the hidden state and an output vector. To elaborate this further, this function accepts the previously hidden state, h_{t-1}, and its weights, W_{hh}, to generate the final output, Y_t.

As you can see, there are three different weight matrices (W_{xh}, W_{hy}, and W_{hy}) in a standard RNN. But why are there three different weights?

First, we have a weight matrix (Wxh) that we use to matrix multiply against the input vector. Similarly, we matrix multiply another weight matrix (Whh) against the previously hidden state, add them together along with the bias term, and pass them through an activation function such as tanh to get non-linearity, as shown in *Figure 6.6*. If we are interested in outputting at every timestamp, then we need another weight matrix (*Why*), which we matrix multiply with the new hidden state before adding the bias term to generate some predictions at every timestamp.

Like FNNs and CNNs, we randomly initialize weights in an RNN, and during training, the network tries to learn the weights against a loss function using backpropagation.

> **Note**
>
> Parameters such as weights matrices and biases are shared across the whole RNN, irrespective of how many timestamps the network have greatly reducing the number of parameters to be estimated.

Since it is very confusing as to what is happening in the RNN block and the feedback loop, let's unroll this RNN block for multiple timestamps to fully understand how an RNN works:

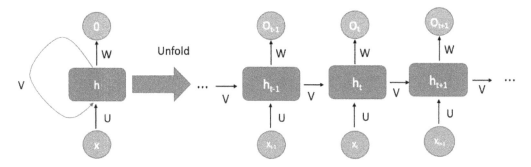

Figure 6.7 – Schematic representation of an RNN block

As you can see in the preceding image, an RNN can be thought of as a series of multiple FNNs, each passing the information from one step to the next:

1. In the first timestamp, the function receives the input, along with weights, and passes them through an activation function such as tanh to generate some predictions.

2. Then, we repeat this process when we receive the next input. So, our input and previously hidden state are passed onto a function and this process repeats multiple times until we finish all the sequences in our input.

3. One thing to note is that we are reusing the same weight matrices at each timestamp, even though the inputs and hidden state values are different. Due to this, a loss is calculated at each timestamp that shows how far we are with predictions using a softmax loss function. Then, we can have a final loss that combines all the individual losses.

A RNN will also adjust the weights using backpropagation which is slightly different compared to FNNs. Let's find how next.

Backpropagation in RNNs

In an FNN, during training, the model weights are updated using backpropagation which is an algorithm used for calculating the gradient of a loss function with respect to the weights of the network. But in the case of an RNN, this is done using **backpropagation through time** (**BPTT**). Backpropagation is used to compute the gradient for the loss function with respect to the weights. Then, all the losses are combined to calculate the final loss (*Figure .6.8*). During BPTT, the RNN receives the derivative of loss and then the error is backpropagated from the last timestamp to the first timestamp through the network to update the weights. As you can see, RNNs differ from standard neural networks in that they use the same weights and biases throughout the network, thereby lowering the number of parameters to deal with. This significantly reduces the complexity and thereby increases the efficiency of the network:

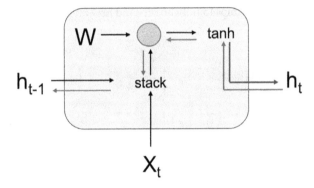

Figure 6.8 – A simplified example of backpropagation in an RNN

One of the problems with BPTT is that it takes a lot of time for the model to train a long sequence and then converge. **Truncated backpropagation through time** (**TBPTT**) can address such issues. The idea is simple: instead of training the entire long sequence at once and then calculating the gradient before doing backpropagation, first, we take steps of, say 10, then compute the loss, then backpropagate, and then perform gradient descent. When we repeat this process, we have hidden states that we computed from the previous chunk, so the forward pass remains the same, but now, when computing the gradients for the second chunk, we will only backpropagate in the second chunk.

Different RNN architectures

Despite the incredible success of RNNs in solving some of the hardest problems in sequential modeling and predictions, the limitation with RNNs is that they put so much emphasis on the most recent inputs. This means that the last input you feed into the network has a higher influence on the prediction compared to the previous timestamps. . When the values of the gradients are too small making model converge slower, this is what we call "memory" loss or the vanishing gradient problem when the values of the gradients are too small making model converge slower. RNNs are not good at remembering long-term associations in the data. Another issue with the vanilla RNN is that it only used information earlier in the sequence. To address the limitations of RNNs, several variations of RNN architectures have been proposed, such as **bidirectional RNN**, **long short-term memory** (**LSTM**), and **gated recurrent unit** (**GRU**). We will briefly look at these popular RNN architectures in the next section.

Bidirectional RNNs (BiLSTM)

In a vanilla RNN with one hidden layer, we pass the input vector at the timestamp. Then, while processing these inputs, the network calculates the hidden state and an optional output vector. They are both dependent on the current input vector and the internal memory. In a deep RNN, there are multiple hidden layers, and each hidden layer uses the internal memory from the previously hidden layer. So far, we have seen how standard RNNs leverage information from the previous hidden state, along with the current state, to make future predictions. However, in some cases, the predictions for input vectors are not only influenced by the previous data but also future data. **BiLSTM** are designed to incorporate this information, scan the data once in forward and once in reverse direction:

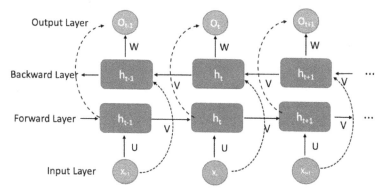

Figure 6.9 – A simple illustration of a bidirectional RNN

As shown in the preceding diagram, the first RNN starts from the beginning and goes to the end of the sequence; then, the second RNN starts from the end of the sentence and goes to the beginning. This way, the model is not limited to the present and past events only and can help with analyzing future events.

LSTMs and GRUs

An LSTM is a type of RNN that is designed for long-term dependencies and addresses the main issue of vanishing gradient problem with RNNs. LSTMs, because of they contain the information in the memory they make RNNs remember inputs over a long time. To address the short-term skipping of events and to remember long-range dependencies in the data, LSTMs use special hidden units. Let's understand the difference between standard RNN and LSTM then:

- Like vanilla RNNs, LSTMs have RNN blocks, but the structure of the cell is very different. LSTMs also have repeating modules, but the repeating module has a different structure (*Figure 6.10*). In vanilla RNNs, you have recurrent modules consisting of an activation function such as tanh to introduce non-linearity, whereas LSTMs use the cell state to decide whether or not to store or delete the information based on the importance it assigns to the information coming into the network, which is carefully regulated by three types of gates – an **input gate**, a **forget gate**, and an **output gate**. All these three gates determine whether or not if the network can take the new input, delete the information since it is not important or effects the output of the current timestamp respectively. Let's understand each of these gates separately now.

- In a typical RNN, the input at the current stamp is fed into the RNN cell, along with the previous hidden state, and is then passed to the activation function to make predictions in the output. However, this process is highly complicated in LSTMs:

    I.   First, the forget gate decides whether the previously stored information from the previous timestamp should be discarded or not (saved).

    II.   Next, the input gate learns new information from the input and adds it to the cell state.

    III. Finally, the output gate takes input from three different gates – the input gate, the forget gate, and the output gate and determine the value of the final output. For more in-depth explanation of how LSTMs work, we encourage readers to check this excellent blogpost by Christopher Olah (`https://colah.github.io/posts/2015-08-Understanding-LSTMs/`).

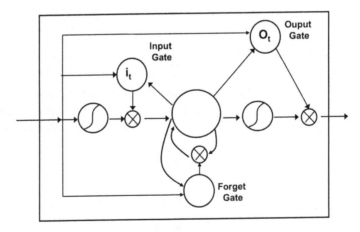

Figure 6.10 - A schematic of a LSTM cell in a RNN

GRUs are like LSTMs except that instead of three gates, GRUs have two – an **update gate** and a **reset gate**. The update gate decides how much previous information gets passed to the next stage in the network. By doing this, it can address the vanishing gradient problem because it can decide to copy all the information into the next state or discard it completely. In contrast, the reset gate decides how much of the previous information needs to be neglected, thereby playing an important role in deciding if the previous information is important or not.

Different types of RNNs

What makes RNNs great compared to FNNs and CNNs is they are not constrained by the size of inputs and outputs. The inputs and outputs can be of different lengths. As we saw earlier, a CNN takes an image of a fixed size as an input vector and predicts an output of probabilities of a fixed size. In addition, the computation is done by using a fixed number of layers. However, RNNs are not constrained by input or output size. Broadly, there are four different input/output combinations with RNNs and each of these types has an application:

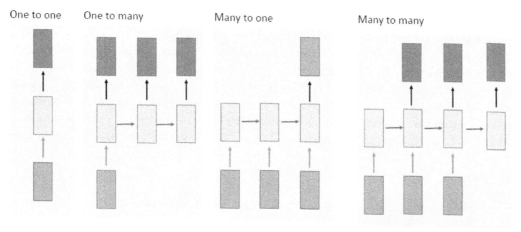

Figure 6.11 – Types of RNNs

Here, each orange rectangle represents an input vector, each blue rectangle represents an output vector, and each green rectangle represents a hidden state. Each rectangle is a vector and the arrow represents a function:

1. The first type of RNN is one-to-one. This is similar to a standard FNN and has a fixed-sized input and an output (For example, DNA classification where the input is an sequence of fixed length say 101 nucleotides and the output is a 2 TF or not-TF). It is not useful for problems where the input and output sizes vary.

2. The second type is one-to-many. Here, we have a fixed-sized input and a non-fixed output (for example, in image captioning, there's a fixed-sized input but a variable number of words to describe the image). There will be output at each timestamp.

3. The third type is many-to-one. Here, we have a fixed-sized output but a variable input vector (for example, in sentiment analysis, irrespective of the size of the input vector, the output is one of the three sentiment types: positive, negative, or neutral. There will be a final output that will help us make a decision based on the final decision of this network. This is because the final hidden state summarizes the context from all hidden states.

4. The fourth type is many-to-many. There are two subtypes:

    A. In the first subtype, the input and output variable sizes do not match (for example, in language translation, which involves passing one or many words in English as input and converting them into one or many words in French as output).

    B. In the second subtype, the input and output have a fixed size (for example, the labels of video frames, where the number of frames in the video matches the labels). For this subtype, there will be outputs at each timestamp.

Applications and use cases of RNNs in genomics

Even though FNNs and CNNs are extremely popular for tackling problems in genomics, they both have limitations. Genomics is all about sequence data, so RNNs can play a key role in several genomics applications. In addition, RNNs can find long-range dependencies in the data, which is why RNNs are great for genomic applications. RNNs are currently being used in several genomics applications, such as constructing a genotype imputation and phenotype sequences prediction system, base-calling accuracy for nanopore sequencing data, genetic regulatory networks, predicting protein functions, and so on. We'll quickly look at a few RNN-based applications in genomics in the following section.

DeepNano

DeepNano is a freely available base caller for the **Oxford Nanopore Technology (ONT)** sequencing platform (https://journals.plos.org/plosone/article?id=10.1371/journal.pone.0178751). Base calling is particularly important for sequencing platforms such as ONT sequencing instruments because they produce exceedingly long reads and a higher error rate. **Hidden Markov Models (HMMs)** are currently being used for base-calling purposes. However, HMMs are not good for long-range dependencies which are also important for accurate base calling. DeepNano is based on deep RNNs and was built to improve the accuracy of the base calling of ONT sequencing data; it performs better than state-of-the-art HMMs. DeepNano is based on GRUs; it processes sequences from the instrument separately and the output layer predicts the base, which is directly used as a base call for downstream applications. Let's understand how the basecalling algorithm works in DeepNano:

- The events in the Nanopore sequence are the context shifts and the changes in the electric current when the DNA moves through the pores in the sequencing instrument.

- Each event represents a summary of the different statistics such as mean, standard deviation, and duration.

- The goal of DeepNano is to take these events as inputs and convert them into a DNA sequence. The output from DeepNano is the probability distribution of base calls. In the case of DeepNano, the predictions are not only influenced by the previous data but also the future data. Due to this, it uses a bidirectional GRU.

ProLanGo

In biology, predicting protein function is one of the important problem. Understanding the role of each of the proteins would not only help us understand their role in our cells but also have tremendous applications in pharmaceuticals, medicine, agriculture, and so on. Experimentally determining the function of the proteins is incredibly challenging, so researchers heavily rely on computational methods to infer the function of the protein while looking at the protein sequence. Several computational methods such as BLAST, GO, protein-protein interactions, gene-gene interaction network analysis, and others are currently being used to identify protein function. More recently, ML and DL-based methods have been more prominent for predicting protein function. These methods are information methods that directly use protein sequences without doing any database searches. Both these use extracted features from the protein sequence, such as the sequence of the protein, the protein's secondary structure, hydrophobicity, and so on, to train and build models and use those models to predict which functional categories a particular protein belongs to. Some popular ML algorithms include KNN, naive Bayes, and support vector machine (SVM). ProLanGo is a novel DL method for deciphering the function of the protein. (`https://www.ncbi.nlm.nih.gov/pmc/articles/PMC6151571/`). It uses protein sequence language called "ProLan" and a protein function language called "GoLan." to convert a protein function problem into a language translation problem. It uses RNN to predict the function of the protein from it's sequence.

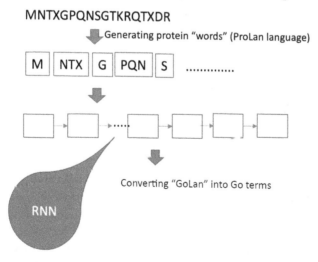

Figure 6.12 – The flowchart of ProLanGo for protein function prediction

As we can see, the sequence of a protein is treated as a words in a sentence. All 20 letters are used to represent the 20 amino acids in a protein. From the protein sequence, using k-mer as protein words, a set of protein words are extracted. This method is called "ProLan," which involves converting protein sequences into protein words based on the frequency of k-mers.

DanQ

Whereas ProLanGo tries to address the challenges with protein function prediction, DanQ (`https://academic.oup.com/nar/article/44/11/e107/2468300`) is a hybrid DL architecture that combines a CNN and a bidirectional LSTM to predict non-coding functions from a DNA sequence (*Figure 6.13*):

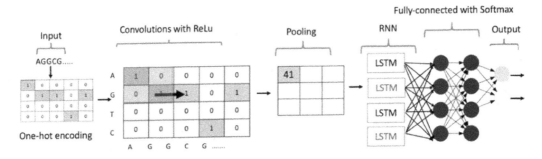

Figure 6.13 – Simplified architecture of DanQ

As shown in the preceding figure, one-hot encoding of DNA sequences is used as an input to the network. A CNN is used first, followed by a Bidirectional RNN with LSTM units. Finally, the outputs from the Bidirectional RNN are fed into fully-connected networks to make final predictions. In DanQ, the upstream component (that is, the convolutional layer) captures features such as regulatory motifs, whereas the later component (that is, the RNN) captures the long-term dependencies between the motifs to learn a regulatory pattern and improve predictions. Compared to previous methods, DanQ achieved more than 50% improvements in performance metrics such as AUC and overall prediction accuracy.

DeepTarget

DeepTarget (`https://arxiv.org/abs/1603.09123`) is an end-to-end algorithm for microRNA target predictions based on the interactions between microRNAs and mRNAs. It uses an RNN-based autoencoder that learns sequence features separately for microRNAs and mRNAs. MicroRNAs are small RNA sequences of size <50 nucleotides and play a key role in post-transcriptional gene regulation by inhibiting translation or degrading their target genes by actively participating in several biological processes. So, identifying target genes for microRNAs is important since the targets are used for downregulation and functional analysis. Target gene prediction of microRNAs is a well-studied computational problem, and several computational and bioinformatics methods have been developed. However, all these methods either rely on manually extracted features of known microRNA and mRNA pairs or fail to filter out false positives. DeepTarget was proposed to model microRNA and mRNA sequences to predict microRNA targets in an unsupervised way. It does so by leveraging two of the most important neural network architectures – autoencoders and deep RNNs:

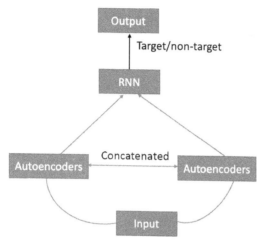

Figure 6.14 – Overview of the DeepTarget architecture

As shown in *Figure 6.14*, the proposed methodology for DeepTarget consists of an input layer that's connected to the first layer, which consists of two autoencoders (an unsupervised DNN that learns latent representation of features; we will learn more about that in the next chapter) that are parallel to the model's microRNA and mRNA sequences. The next layer is the RNN layer, which models the sequence-to-sequence interactions between the microRNA and target mRNA sequences. Finally, RNN's is connected to a fully dense output layer with two nodes to perform binary classification on the target and non-target sequences. The algorithm uses unsupervised feature learning and autoencoders to predict the microRNA target without any predefined features. Now that we have looked at the RNN and its architecture, let's understand how an RNN works on a real-world genomics problem.

Understanding RNNs through Transcription Factor Binding Site (TFBS) predictions

Transcription factors (TF) play a key role in gene regulation, particularly during transcription, where they bind to the promoter regions and initiate the process of transcription. **Transcription Factor Binding Sites (TFBSs)** in DNA are short sequences in gene regulatory regions (such as promoters) and typically range in size from 5 bp to 20 bp. Each TF binds to a different TFBS and controls gene regulation in the cell. Thus, identifying the TF binding sites is key for us to understand cellular and molecular processes. Several experimental methods can identify TFBSs, such as ChIP-Seq technologies and databases such as ENCODE, which have made TFBS information available to researchers. However, ChIP-Seq technologies are expensive, slow, and laborious, and cannot find patterns in the identified

TFBS. Several computational methods have become the go-to for solving this very important problem of identifying TFBSs. Given a particular sequence, predicting whether it is a TFBS or not is the core task of bioinformatics.

In the following toy example, let's see how we can use RNNs to predict a TFBS from DNA sequences. The problem of TFBS can be thought of as a binary classification problem – that is, whether the TFBS can be found in a DNA sequence or not, which we represent as 1 or 0, respectively. The input to the RNN model is the input DNA sequences and their targets, which have labels of 1 or 0. The goal here is to build a highly accurate classification model using an RNN that can be used to derive consistent sequence features from many genomic datasets. The TFBS training data consists of DNA sequences with and without TFBSs, as shown in *Figure 6.15*:

Sample 1:

aggcagtagagccgaggcgcaggtgcacactagacagggcccgcccgcgccGCCCCGCCCCTggcggtagctagcatcgccccgc
cttcccatccggagcgtgacgtcaaggggg

Sample 2:

cgggaaagatagcaggatgtggaggcgatctgggggcgggtcaaacgctgGCCCCGCCCCTctgcaggcccctacctatcagtgtc
catacgtgcttgtgtagacagcgcctg

Sample 3:

agaggcgagtgatcggacaacttcgtttgacagtacggaagcggctgcgccgGCCCCGCCCCTccagaggcaggtgagttagtggc
ctcatgaataatgaatcgtgcaggagaggag

Sample 4:

agatgatgagacccggtagagatagatagccagtagagggaaatttgacgatgagagcagatagaccagttagcagatagacgatagacagat
acagatagacagatagccatagacagagtaga

Sample 5:

tcagttagagacgatagacagatacgatcagtggaagtagaccagataggacccagtagccgataggcagatagccagtagggtttagcgatg
aacgatagcagatagcagatagccgatgacgat

Figure 6.15 – Example DNA sequences showing positive and negative TFBS

As shown in *Figure 6.15*, samples 1 to 3 consist of positive TFBSs (label=1), where the DNA sequences consist of TFBSs, whereas samples 4 and 5 are negative (label=0):

- The first thing we must do is convert DNA sequence into a one-hot encoding vector. To refresh your memory, a one-hot encoding vector converts each nucleotide of the DNA sequence into a binary vector, labeled 0 or 1.

- After the input is fed into the RNN, it produces another matrix. As we just learned, at each timestamp, the RNN takes an input vector and the previously hidden state vector and produces the new hidden state recursively. In this case, each position in the sequence is a timestamp. At the end of the training process, the RNN produces an output vector at the timestamp of the input sequence (*Figure 6.16*).

- Then, the output vector from the RNN is fed into a softmax activation function in the last layer of the network, which learns the mapping between the hidden space and the target label (0 or 1). The final output is a probability that indicates whether the DNA sequence is a TFBS or a non-TFBS.

- Like other FNNs and RNNs, we calculate the loss (cross-entropy loss) and then the model is trained until the network generates low or no loss. This minimization of the loss function is achieved using the BPTT algorithm. We can use dropout as a regularization method for the model to prevent overfitting:

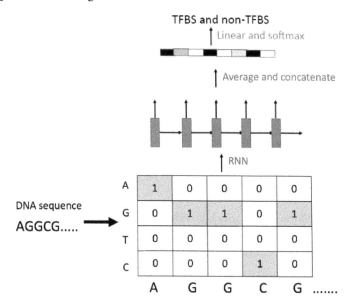

Figure 6.16 – Illustration of an RNN for the TFBS task

After the model is trained and evaluated on test data (which you will learn in *Chapter 9, Building and Tuning Deep Learning Models*), then you can use that trained model and predict if a new DNA sequence has a TFBS or not.

TFBS prediction via a ChIP-seq experiment

Now that we have understood how RNNs can be used on a toy example, let's use our knowledge to predict TFBSs via a real-world ChIP-seq experiment.

Data collection

The dataset that we are going to use to understand BiLSTM consists of 690 ChIP-Seq experiments. The dataset has already been preprocessed, so we don't need to perform any preprocessing. Since the original dataset is huge but for the current hands-on tutorial, we will only take a subset of these samples.

Data preprocessing

The dataset has already been preprocessed and represented as a 1,000 x 4 one-hot encoded matrix. The target dataset consist of 1 or 0.

Let's understand how the input dataset was generated:

- The human reference genome was first divided into 200 bp non-overlapping segments
- For each of the 690 ChIP-seq experiments, if 100 bp-200 bp segments belonged to a peak, it was classified as positive (label=1); otherwise, it was classified as negative (label=0)
- 800 bp (400 bp on either side) sequences was addede to both sides of the 200 bp segment to create a 1,000 bp input sequence

Data splitting

The data, which consists of 4,863,024 samples, was split into three different datasets in the ratio of 90:0.1:9.36 for training, validation, and test, respectively. The training data consists of a total of 4,400,000 samples, the validation data consists of 8,000 samples, and the test data consists of 455,024 samples.

Model training

To train this dataset, we will use LSTM, which is widely used in genomics applications due to the use of sequential data, such as DNA sequences. We will use BiLSTM to classify ChIP-seq data. Let's go through the architecture that we will be using to train the data:

- After feeding input to the network through the input layer, the next layer will be a CNN. You might be wondering why we are using a CNN since the goal is to leverage an RNN for a genomics problem. This is because the CNN layer acts as a motif scanner, as we learned in the previous chapter.
- The output from the CNN is fed into the BiLSTM layer. The output from the BiLSTM layer is then flattened and fed into a fully connected layer.
- In the output layer of the network, a sigmoid function is applied. The final output is a 690-dimensional vector, where each element corresponds to the ChIP-seq experiment.

Now that we understood the architecture of how we want to build a RNN to predict the TFBS for a DNA sequence, let's use Keras library to build the network.

1. First, let's load the necessary libraries:

    ```
    import numpy as np
    from sklearn import metrics
    import pandas as pd
    from keras.models import Sequential
    ```

```
from keras.models import Model
from keras.layers import Dense, Dropout, Activation,
Flatten, Layer, Input
from keras.layers.convolutional import Conv1D,
MaxPooling1D
from keras.layers import LSTM
from keras.layers import Bidirectional
from keras.callbacks import ModelCheckpoint, EarlyStopping
```

2. Now, load the training data:

```
# Load train labels data
y_train = np.load('data/y_train.npy')
y_train = y_train[:10000,:]
# Load train labels data
y_train = np.load('data/y_train.npy')
y_train = y_train[:10000,:]
```

Here, we have loaded the training and validation data from the downloaded file.

3. Now that the training and validation data is ready, we can compile the CNN-BiLSTM model and start training the model:

```
# Load train labels data
y_train = np.load('data/y_train.npy')
y_train = y_train[:10000,:]


# Load train labels data
y_train = np.load('data/y_train.npy')
y_train = y_train[:10000,:]
y_train.shape


input_data = Input(shape=(1000,4))


# Convolutional Layer
output = Conv1D(320,kernel_size=26,activation="relu")
(input_data)
output = MaxPooling1D()(output)
output = Dropout(0.2)(output)
```

```
#BiLSTM Layer
output = Bidirectional(LSTM(320,return_sequences=True))
(output)
output = Dropout(0.5)(output)


flat_output = Flatten()(output)


#FC Layer
FC_output = Dense(695)(flat_output)
FC_output = Activation('relu')(FC_output)


#Output Layer
output = Dense(690)(FC_output)
output = Activation('sigmoid')(output)


model = Model(inputs=input_data, outputs=output)


print('compiling model')
model.compile(loss='binary_crossentropy',
optimizer='adam')


print('model summary')
model.summary()


checkpointer = ModelCheckpoint(filepath="./model/bilstm_
model.hdf5", verbose=1, save_best_only=False)
earlystopper = EarlyStopping(monitor='val_loss',
patience=10, verbose=1)


history = model.fit(X_train, y_train, batch_size=100,
epochs=2, shuffle=True, verbose=1, validation_split=0.1,
callbacks=[checkpointer,earlystopper])
```

4. Let's look at the training loss and validation loss curve and see how well we have done regarding the model training:

```
np.mean(history.history['loss'])
0.07372421585023403
```

```
np.mean(history.history['val_loss'])
0.05934176221489906
```

It's not too bad considering we have only ran for 2 epochs and haven't tried adjusting the parameters.

As an exercise, you are encouraged to take the model that you have built here and see how well it generalizes on the test data that is available in the downloaded folder. Do you think anything can be improved?

> **Note**
> If you are stuck or need some help, please refer to the solution in here (`https://github.com/PacktPublishing/Deep-Learning-for-Genomics-/blob/main/Chapter06/test.ipynb`)

Summary

RNNs are a special type of neural network that is well suited for sequential data such as time series, audio, video, and text. Research showed that RNNs have improved the performance of sequential data types when compared to other architectures such as FNNs and CNNs. The key to an RNN is the sequence memory state, which helps it store information from the previously analyzed state; this is good for sequential signal analysis and predictive analysis. In this chapter, we learned how RNNs are different from FNNs and CNNs. We understood the different types of RNNs and what makes them good for sequential data analysis by looking at a few examples. RNNs, as you may have noticed, are good for mapping a fixed or variable-sized input sequence to a fixed or variable-sized output; we have seen several examples to understand this.

We also looked at how RNNs can help with genomics tasks and understood the different architectural types of RNNs. Bidirectional RNN, LSTM, and GRU are variants of RNNs that are capable of long-term associations, thereby retaining the information from an infinite sequence, which is very common in genomics. They address long-term dependencies.

You were also introduced to the different RNN types and their applications in various domains, such as image captioning, language translation, and others. Finally, we looked at how RNNs are used to solve some of the key problems in genomics, such as TF binding site detection, miRNA-mRNA sequence modeling, gene expression analysis, histone modifications, base calling, and more. In the next chapter, we will look at another exciting neural network architecture called autoencoders, which has a lot of potential applications in genomics.

7
Unsupervised Deep Learning with Autoencoders

Over the past few years, data-driven **deep learning** (**DL**) approaches have made impressive progress in the genomics field. The development of high-throughput technologies such as **next-generation sequencing** (**NGS**) has played a major part in this data-driven revolution. Several **neural network** (**NN**) architectures have found success in the genomics domain. For instance, in the previous chapters, we have seen **feed-forward neural networks** (**FNNs**), **convolutional neural networks** (**CNNs**), and **recurrent neural networks** (**RNNs**), which have been successfully used for many genomics applications. So far, all these NN architectures require that you have well-labeled data. However, a lack of ground truth and accurate labels is common in the genomics domain, which limits the application of **supervised learning** (**SL**) methods. NGS has significantly increased the use of gene expression assays, and there is so much genomics data out there with no label. Several methods exist to generate signals from these unlabeled datasets such as **clustering**, **principal component analysis** (**PCA**), **multidimensional scaling** (**MDS**), and so on, but integrative analysis involving multi-omics datasets is challenging with these conventional methods. Furthermore, in non-model species, the genome-wide gene expression analysis is very challenging because of the unavailability of gene function information and lack of knowledge of the organism's biology. What we need are unbiased approaches that can find patterns in these large and complex unlabeled datasets in an unsupervised way.

Autoencoders are a type of NN architecture that can harness the power of DL in an unsupervised way. They are the most important NN architecture for **unsupervised learning** (**UL**). Unlike vanilla NN architectures, which model data by minimizing the loss between the predictions and original data, autoencoders try to learn an identity function that minimizes the loss function. Autoencoders have found a lot of success in several applications such as image compression, image denoising, image generation, recommendation system, **sequence-to-sequence** (**seq2seq**) prediction, and so on. In this chapter, you will understand what unsupervised DL is, why it is important, the different types of unsupervised DL methods, what autoencoders are, and, finally, the different applications of autoencoders in genomics. By the end of this chapter, you will know what unsupervised DL is, how autoencoders work, and the genomics applications of autoencoders.

As such, the following is an outline of this chapter:

- What is unsupervised DL?

- Types of unsupervised DL

- What are autoencoders?

- Autoencoders for genomics

What is unsupervised DL?

Unless you are lucky, the chances are that most of the data that comes to you is unlabeled, whether it is images on the web, text from a document, gene expression data from NGS experiments, and so on. Even if they come labeled, they are not clean and perfect datasets. This is where UL algorithms are useful. In UL, the algorithm is presented with the training datasets without any label, which means these datasets don't have a particular outcome or specific instructions on what to do with them. The job of the UL model is to automatically extract features from unlabeled datasets and use those features to find hidden patterns. The unsupervised models first try to extract simple features from the data, then stitch them together to form more advanced features, and finally, come up with an outcome. Unlike SL models, these models don't have a ground truth to evaluate the performance of the models using metrics such as accuracy, **mean squared error** (**MSE**), AUC score, AUC-ROC score, and so on. Despite these limitations, UL is very important since getting labeled data is either impractical or expensive, and unsupervised models can find structure in the data and produce high-quality results.

Now that we have some background on unsupervised DL, let's dive deep into the concepts and applications of unsupervised DL for genomics.

Types of unsupervised DL

There are broadly three different types of UL methods that are currently available:

- Clustering

- Anomaly detection

- Association

Let's discuss each of these in detail in the following sections.

Clustering

Clustering, as the name suggests, is a type of UL method to group similar data points in the training dataset—for example, the clustering of tissues based on the gene expression values from genomics data. This is the most common method of UL. Here, the DL models look for similar data points in the training data to group them using the appropriate distance measurement method (*Figure 7.1*).

One challenge with the clustering method is you need to predefine the number of clusters for the algorithm to group clustering based on the number of clusters. However, there are methods out there that can help arrive at this cluster size to input into the learning algorithm.

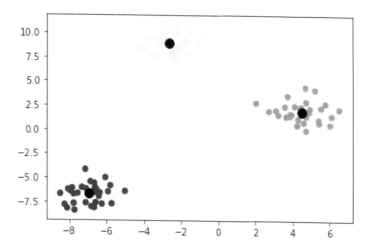

Figure 7.1 – Clustering of multiple samples into definite groups

The preceding figure shows the clustering of multiple samples into three different clusters.

Anomaly detection

This is a type of UL that is mainly used for detecting abnormal or unusual behavior as compared to an established behavior, hence the name *anomaly detection*. Anything that is deviating from the established baseline behavior is called an anomaly. Among anomaly detection methods, the most important is outlier detection, followed by the novelty detection method.

Outlier detection

Let's understand what an outlier is first. Outliers are a fraction of samples that differ from the rest of the samples in the population. There are several reasons why this happens—for example, in experimental procedures, those samples are functionally different from the majority of the samples, and so on. In general, **exploratory data analysis (EDA)** and data processing methods can help remove outliers that arise because of experimental procedures so affect downstream analysis. However, in the latter case, those outliers need to be further analyzed to find the exact reason why they are different from the rest of the samples. In biology and genomics disciplines, outlier detection is very important because it may indicate an important event that is far separated from the rest of the other data points, and so a closer inspection may reveal some interesting insights. For example, a few genes may unusually show higher expression compared to the rest of the genes, which might indicate that these might be relevant to some biological condition.

As can be seen in the following screenshot, running the data with an outlier detection method has discarded several outliers:

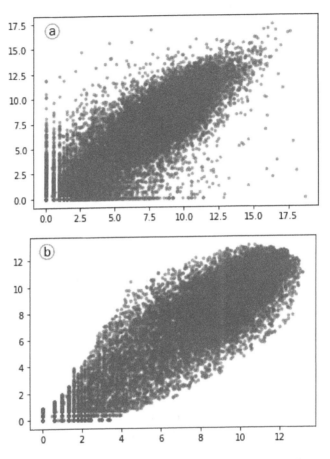

Figure 7.2 – Outlier detection before and after removing outliers

The preceding shows how the data looks with outliers (a) and after removing outliers (b).

Novelty detection

In contrast to outlier detection, the original data does not contain any outliers. Here, we are interested in identifying whether a new observation is an outlier, which we refer to as a novelty. Let's understand this bit further. Consider a dataset of n samples coming from the same distribution. Suppose we add one more observation to that dataset. There are two possibilities—is the new observation so different from the rest of the observations that it is not considered regular (novel) or is the new observation so similar to the other observations that we cannot distinguish it from the other observations? This is where novelty detection methods are useful.

Association

Association is where the algorithm tries to learn new and interesting insights hidden from the training dataset without any labels. The relationship is returned in the form of rules or frequent item sets. The unsupervised algorithm finds certain key attributes (features) of the data sample and tries to predict other attributes with which they are closely associated. One thing to remember with this method is the association rule is a descriptive, not a predictive, method.

Now that we understand what UL approaches are, let's now dive deep into understanding autoencoders, which are one of the two most popular unsupervised DL approaches that can harness the power of DL without the need for labels in the data.

What are autoencoders?

Autoencoders are a type of **deep NN** (**DNN**) that can learn an efficient reduced representation of the data in an unsupervised way and minimize the error between the compressed and subsequently reconstructed data compared to the original data. Why compress the data and then reconstruct the original data? Isn't it counterintuitive? Suppose you are on your vacation and took pictures, but you realized after you return from vacation that a picture has noise because of dim light. Wouldn't it be nice if there was a way to remove the background and make the picture great? This is, in computer vision lingo, called **feature variation**, which removes any noise in a picture. This is what autoencoders do. They learn a representation or latent space from the training data to ignore signal *noise*. The compression step forces the network to only learn the most important latent features. This is because if the model is at full capacity, it will just copy the data without learning any useful features regarding the data structure. During reconstruction, the network ignores the non-essential features and thereby ignores signal *noise*.

Autoencoders compress the data into a reduced representation of a low-dimensional latent space using multiple non-linear transformations through layers in a DNN. This concept may sound familiar to you because we have a similar method called **PCA** that does the same thing. So, what's the difference between the two different methods? Well, the biggest difference between PCA and autoencoders is the type of transformation. Let's understand it from the following outline:

- PCA is a **dimensionality reduction technique** (**DRT**) that converts the input values into a few components that best represent the variance in the data and is mainly a linear transformation method.

- An autoencoder, on the other hand, is a non-linear transformation method that uses a non-linear activation function and multiple NN layers to transform the data and compress it. Autoencoders are more efficient to learn model parameters because they use several layers, unlike PCA, which can learn from one huge transformation. Autoencoders can make use of pre-trained models that were trained and optimized on different datasets (a process termed **transfer learning** or **TL**), whereas PCA needs to be run every time on a new dataset.

> **Note**
>
> One thing to note is that the autoencoder doesn't need to be a fully connected DNN as seen in vanilla NNs but can also be convolutional layers that best work for audio, video, and sequential data.

Autoencoders are quite useful for non-linear transformation, which is a common task in many datasets, such as sequential, image, text, genomics datasets, and so on. Let's understand some of the properties of autoencoders in the next section.

Properties of autoencoders

The three main properties of autoencoders are 1) they are data-specific, 2) they are lossy, and 3) they learn automatically instead of being manually engineered by humans. Let's look at this in more detail:

- Autoencoders only work on the data on which they have been trained which means their behavior is data specific

- Compared to the original input, the output from Autoencoders is always degraded and hence it is lossy

- Similar to other DL models, Autoencoders learn features from the data automatically and so there is no need to extract features manually.

How do autoencoders work?

So, why do autoencoders reconstruct the input as an output? Why is it of interest to us, then? The trick is, unlike other supervised NNs that we have seen so far, we do not care about the output here, but what we do care about is the compressed representation (or 'latent space') of the data in the hidden layers. If we reduce the number of neurons in the hidden layers, then the network tries to learn the weights and biases with few neurons in the hidden layers to create a dense representation of data (*Figure 7.3*). That means the hidden layers in the autoencoder only have to learn the most key features of the input data. For example, if we train an autoencoder on a genome without any mutations (reference genome or gold standard reference), it will learn mutations as something abnormal when used on another genome. This is exactly what an autoencoder does.

You can see a visual representation of Autoencoder in here:

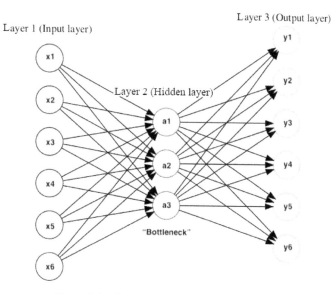

Figure 7.3 – Compressed representation of data

As portrayed in *Figure 7.4*, we take an unlabeled dataset and convert it to an SL dataset by generating outputs as shown in the output layer, which is reconstructed from the original input. During training, we minimize the reconstruction error, by taking the difference between the original input and the reconstructed output. The output is the same as the input, so we do not need to create labels, hence an autoencoder is good for unlabeled data:

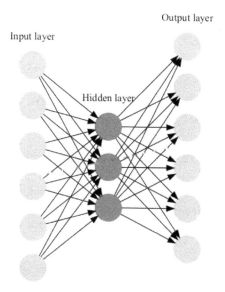

Figure 7.4 – Basic concept of an autoencoder

So, in summary, an autoencoder does three main things:

- Takes the input data
- Compresses the input and learns a representation
- Reconstructs the output without noise and background

That's it. Now, let us understand the architecture that allows it to do these tasks effectively.

Architecture of autoencoders

The overall goal of autoencoders is to derive an identity function in an unsupervised way so that it can reconstruct the original input through a reduced representation of the input.

Have a look at the following diagram:

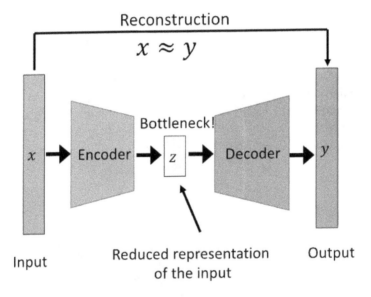

Figure 7.5 – Architecture of an autoencoder

Figure 7.5 shows a typical autoencoder structure where x is the input, z is the reduced representation of the input, and y is the output. The rectangles represent layers in a NN architecture. Autoencoders try to reconstruct the original input, and so the reconstruction is $x \approx y$.

So, how does an autoencoder ensure that it reconstructs the original output? Let's look at the basic architecture of an autoencoder to understand more about it. Leaving out input and the reconstructed output, an autoencoder has three main blocks:

- **Encoder**: The encoder is the part of the network that is mainly responsible for compressing the input into latent representation.

- **Bottleneck**: The bottleneck layer or latent or reduced representation layer is a reduced dimension and is the most important attribute of an autoencoder. This hidden layer takes the input data and compresses that within the bottleneck layer. It constrains the amount of information, enabling learned representation of super-complex data in a non-linear way in only a few different dimensions. Without a bottleneck, the network would just simply memorize the input values by passing through the network.

- **Decoder**: The decoder is the part of the network that is finally responsible for reconstructing the output. It evaluates the representation of the data and provides feedback to the network.

In addition to the preceding three fundamental blocks, other key architectural components of an autoencoder are the *loss function* and *regularization*. An autoencoder uses the loss function to compute the error between the original inputs and reconstructed output, and this way, the loss function encourages the model to be sensitive to the inputs. The next term is regularization, which prevents the model from overfitting. Let's understand how these three blocks work, using a couple of examples.

Image compression

Image compression is a very typical application of an autoencoder. Let's have a look at this in more detail:

- In this case, for example, if we pass an image of dimensions (*32,32,1*) to the input layer, we need to flatten the image before we can pass it to the NN architecture.

- A flattened image now becomes 1,024 (32 x 32 = 1,024) and this is then passed to the encoder block. The encoder compresses the input.

- The output from this encoder block is then fed into the bottleneck layer, which consists of 8–16 neurons.

- Next, the decoder block tries to reconstruct the original image from the bottleneck layer, as shown in *Figure 7.6*.

- If you look at the input and reconstructed output images, they both look the same, so you might be wondering how you would know whether we have successfully reconstructed them or not. This is where we use the loss function, which computes the difference between the input and output images, which can then be minimized as shown in the following figure:

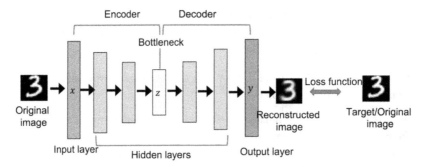

Figure 7.6 – How an autoencoder works for image compression

How can we leverage autoencoders for genomics applications? Now, let's look at a genomics example. For this, we will use **Single-cell RNA sequencing (scRNA-seq)**, which is a very popular sequencing technology for single-cell sequencing.

scRNA-seq

In the simplest use case, we want the autoencoder to take input data, learn a representation of the data in an unsupervised way, and output the data without noise or reconstruct the input as output. Let's see an example of how it can be done in genomics. In a scRNA-seq technology, zero counts can arise either because of low mRNA capture rate (**dropouts**), due to experimental limitations, or because mRNAs are not expressed in that cell type (**structural zeros**). The main challenge while doing scRNA-seq data analysis is deciding which genes are zero counts due to dropouts versus structural zeros. Several methods exist to address this issue, such as imputation, which relies on correlation structure between the cells and genes to differentiate dropouts from structural zeros. We can treat this as a denoising problem and try to denoise the gene expression data to remove any background noise (dropouts) so that only structural zeros are present.

In the following diagram, we have a typical autoencoder with encoder, bottleneck, and output blocks:

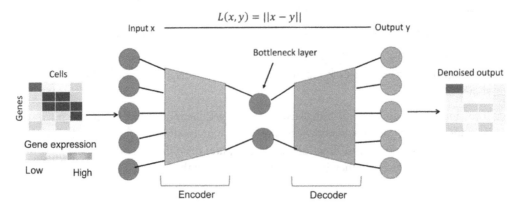

Figure 7.7 – Autoencoders for denoising gene expression data

The input as shown above is a gene expression matrix consisting of cells as columns and genes as rows. In this example, we have five genes, and each gene has expression values from each of the five cell types. The dark-colored boxes indicate zero counts. During the training of an autoencoder, we pass this gene expression matrix in the input layer (five nodes here represent five genes), which are then passed to the **Encoder** block. The output from the **Encoder** block is the reduced representation in the bottleneck layer, and finally, the decoder reconstructs the original input from bottleneck layer. The output layer also has a similar number of nodes to the input (five), and it may contain values of the mean of the negative binomial distribution and the final output matrix that shows the denoised expression.

In this way, autoencoders compress the input into low-dimension space and learn the features such as similarities of non-zeros there to decide how to fill in the blanks or denoise the data, as shown in this example.

Types of autoencoders

In general, there are six types of autoencoders. Let's briefly discuss each of those in the following section.

Vanilla autoencoders

A **vanilla autoencoder** is also called a complete autoencoder or a standard autoencoder that we have seen before. This is a very simple autoencoder that has three layers—an input layer, a bottleneck layer, and an output layer. Here, the input and output are the same, and reconstruction of the original output is learned using an optimizer such as Adam. We can use some loss functions such as MSE. The input might be an image of size 28 x 28 x 1 pixel, which will be n=784 nodes, and the output will be the same size, so it will be n=784 nodes also. The bottleneck layer can be n=64 nodes, which is smaller than the input size of 784, which will help the autoencoder to learn the compressed representation of the data. By constraining the number of nodes in the bottleneck layer, we can restrict the amount of information that can be passed through the network. The model is trained until the reconstruction loss is minimal. Since there is no regularization term in this type of autoencoder, we must ensure the bottleneck layer has a restricted number of nodes so that the network is not copying the original input into the output layer.

Deep autoencoders

Deep autoencoders, in contrast, have multiple layers in both the encoder and decoder blocks but just have one bottleneck layer. In this type of autoencoder, we use a stack of layers.

Convolutional autoencoders

In *Chapter 5, Introducing Convolutional Neural Networks for Genomics*, you learned about convolutions and how they are used for image recognition, image classification, transcription factor-binding site detection, and so on. If you remember, we used a fully connected layer before the final output layer in that architecture. We can also use autoencoders instead of fully connected layers in a CNN architecture. During training, the image or one-hot encoded DNA sequence is downsampled using several layers in the CNN such as convolution and pooling layers, and having autoencoders instead of a fully connected layer ensures we learn a compressed image or data. In practice, autoencoders used in image settings are always **convolutional autoencoders** because they perform much better than the other types of autoencoders for image data.

Regularized autoencoders

So far, we have seen autoencoders that work mainly by placing constraints in the network so that they can learn the useful structure from the data. Instead of putting constraints on the model capacity by keeping the encoder and decoder shallow and the bottleneck layer small, we can impose sparsity in the hidden layers even if the number of hidden units is larger. So, how do we do that in the network? It is mainly done through regularization (such as in SL). There are three different types of **regularized autoencoders**—sparse autoencoders, contractive autoencoders, and denoising autoencoders.

Sparse autoencoders

A sparse autoencoder is a type of regularized autoencoder that doesn't put any constraints on the number of nodes in the hidden layer. Instead, it penalizes the activations within a layer. This is different from traditional regularization methods such as Lasso and Ridge where we penalize the weights of the edges. This way, we encourage only a certain number of nodes in the encoder and decoder blocks.

You can see a visual representation of this here:

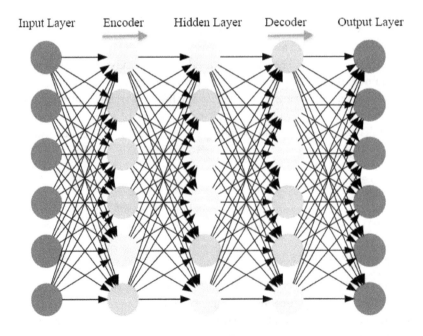

Figure 7.8 – A simplified sparse autoencoder

As can be seen in *Figure 7.8*, only a certain number of nodes are activated in the encoder and decoder block, shown by the pale-colored nodes based on the level of their activation.

Contractive autoencoders

These are similar to regularized autoencoders, which we have seen before, except that the learning is done by applying a penalty term to the loss function. It is better than the other two regularization-based autoencoders for feature learning.

Denoising autoencoders

A denoising autoencoder introduces artificial noise (by converting some inputs into zeros) in the data so that the autoencoder doesn't just copy the data as output without learning any structure of the data. The goal of denoising autoencoders is to recover the undistorted input, as shown in *Figure 7.9*. In this case, the autoencoder learns an approximation mapping function toward a lower dimension that best describes the natural data without any noise. The loss function tries to reduce the output and the noise input. Denoising autoencoders are a very popular type of autoencoder and are mainly used in image denoising to remove any unwanted noise in images and remove dropouts (non-zeros introduced to a low RNA capture rate) in scRNA-seq data. We have seen an example of this in the previous section:

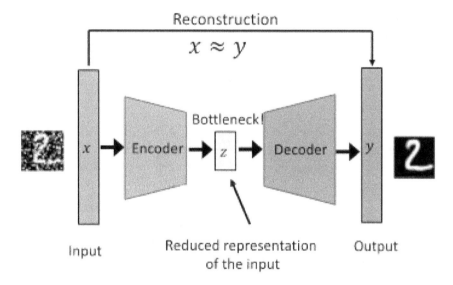

Figure 7.9 – An example of a denoising Auto encoder

The preceding figure shows an example of denoising autoencoder that takes an input image with noise and produces a cleaned output.

Variational autoencoders

Variational autoencoders (VAEs) are modern types of autoencoders. Both VAE and GANs (which we will learn about in the next chapter) learn a latent variable model from the input data. The main difference between a VAE and other autoencoder types is parameters of a probability distribution of the data. Whereas typical autoencoders are deterministic and try to reduce the error between output and original input (reconstruction error), VAEs, in contrast, are stochastic. Unlike other autoencoders, this is very complex and it is beyond the scope of this chapter to go into the details of a VAE.

Autoencoders for genomics

Several applications of autoencoders for genomics exist. The most common type of application, however, is for predicting gene expression from microarray and RNA-Seq datasets. Let's understand how autoencoders work for gene expression analysis.

Gene expression

The main application of autoencoders, as you learned in the previous section, is for gene expression analysis, which includes

- Time-series gene expression where they are mainly used at the preprocessing step for clustering, cDNA microarrays

- RNA-Seq, where they are used to predict the organization of transcriptomics machinery

- Gene expression, where they are mainly used for identification of biological signals and patterns respectively

In a typical gene expression experiment, the inputs are typically numerical values estimating how much RNA is produced from a DNA template through transcription across various cells, tissues, or conditions. Let's look at some popular autoencoder applications for predicting gene expression.

ADAGE

Analysis using Denoising Autoencoders for Gene Expression (**ADAGE**) (https://www. biorxiv.org/content/10.1101/030650v1) is a popular method for unsupervised gene expression analysis using stacked denoising autoencoders. Briefly, the ADAGE model integrates data without requiring any gene function, gene pathway, or experimental labeling, and so making it a suitable method for any large expression datasets. Briefly, random noise is added to the expression data from various samples. Then, these corrupted expression datasets are processed as inputs to the encoder of the autoencoder NN. The output from the encoder is fed into the bottleneck layer consisting of 30 nodes. The encoder tries to generate the output, and the resulting ADAGE model outputs genomics measurements, which are biological features such as strain, oxygen, and so on.

Learning and representing the hierarchical organization of yeast transcriptomic machinery

In this genomics application, stacked autoencoders are used to learn biological features in yeast from microarray experiments (`https://bmcbioinformatics.biomedcentral.com/articles/10.1186/s12859-015-0852-1`). The hidden or bottleneck layer that was trained from these autoencoders is then used to encode biological representations in yeast.

Boosting gene expression clustering

Autoencoders are currently used to cluster gene expression profiles as a powerful tool for inferring information from unsupervised data such as gene expression profiles (`https://www.biorxiv.org/content/early/2017/11/05/214122`).

Use case – Predicting gene expression from TCGA pan-cancer RNA-Seq data using denoising autoencoders

Gene expression analysis is a typical use case in the genomics domain. Gene expression can reveal a lot of biological insights about the state of the cell, tissue, or condition. Currently, a lot of techniques, both traditional and non-traditional, are available for gene expression predictions. Autoencoders, because of their ability to reconstruct data, were proven successful previously to extract novel biological insights from gene expression data. With that goal in mind, in this current use case, we will train an autoencoder from **The Cancer Genome Atlas** (**TCGA**) pan-cancer RNA-Seq data.

Data collection

TGCA consists of several genomics measurements from several tumors across different cancer types. Among them, gene expression measurements can capture information about the biological state of each tumor and is great dataset for our purposes. For this use case, we will train and evaluate an autoencoder network from this data consisting of the 5,000 most variably expressed genes from 10,459 samples. This whole dataset is referred to as pan-cancer data. The data was downloaded from `https://github.com/greenelab/tybalt/blob/master/data/pancan_scaled_zeroone_rnaseq.tsv.gz`.

Data preprocessing

The pan-cancer data is already preprocessed and batch-corrected, so there is no need to do any further data preparation. In addition, the variability of gene expression was calculated by **median absolute deviation** (**MAD**).

Model selection

For this use case, we will select an denoising autoencoder that accepts the 5,000 genes as input reduced representation of 100 features, and reconstruct to the original 5,000 genes.

Model training

Let's start training a denoising autoencoder. This phase can be divided into multiple sub-phases, which we will see in the following sections. Follow the next steps:

1. Start by importing the libraries first:

    ```
    import os
    import numpy as np
    import pandas as pd
    import matplotlib.pyplot as plt
    import seaborn as sns
    from keras.layers import Dense
    from keras.models import Sequential
    from sklearn.model_selection import train_test_split
    ```

    Here, we are importing all dependencies (libraries and packages) that are needed for autoencoders. We use Keras for creating an autoencoder NN. To do this, we will import the Sequential API, which will allow us to build the network easily.

2. Next, we will import the TCGA RNA-Seq data that we discussed previously:

    ```
    pancan_rnaseq_df = pd.read_csv(pancan_scaled_zeroone_
    rnaseq.tsv, index_col=0, sep="\t", low_memory=False)
    pancan_rnaseq_df.iloc[:2,:5].head()
    ```

    After we download the data from the TCGA portal, we import that data into the workspace using pandas' read_csv function and print the first two rows and five columns. The following table shows the first two rows and five columns of the pancan RNA-Seq datasets. Each row corresponds to the sample name and each column corresponds to each gene and its expression in that sample:

	RPS4Y1	XIST	KRT5	AGR2	CEACAM5
TCGA-02-0047-01	0.678296	0.289910	0.034230	0.0	0.0
TCGA-02-0055-01	0.200633	0.654917	0.181993	0.0	0.0

Figure 7.10 – First two rows and five columns of the pancan RNA-Seq datasets

3. **Train-test split**: By now, you are familiar with this concept. Just to refresh your memory, we need to split the full dataset into training and test datasets. The training dataset will be used for training and validating the model, and the test dataset will be used for model evaluation:

    ```
    pancan_rnaseq_df_train, pancan_rnaseq_df_test = train_
    test_split(pancan_rnaseq_df, test_size=0.1, shuffle=True)
    ```

```
pancan_rnaseq_df_train.shape, pancan_rnaseq_df_test.shape
((9413, 5000), (1046, 5000))
```

Here, we are splitting the data in the ratio of 90:10 train:test. The final data dimensions are also printed to confirm that the data was split properly.

4. **Defining the layers of the autoencoder**: Next, we will define the layers of our autoencoder NN. For this, we will use the `Sequential` API of Keras, which is simpler but less flexible compared to the *Functional* API that you have learned before. The code is illustrated here:

```
# This is the size of our encoded representations
encoding_dim = 100
numb_of_features = pancan_rnaseq_df.shape[1]
# Defining the 'Autoencoder' full model
autoencoder = Sequential()
autoencoder.add(Dense(encoding_dim, activation="relu",
input_shape=(numb_of_features, )))
autoencoder.add(Dense(numb_of_features,
activation="sigmoid"))
autoencoder.summary()
Model: "Autoencoder"
_____

 Layer (type) Output Shape Param # ========================
=============================================
 dense_6 (Dense) (None, 100) 500100
 dense_7 (Dense) (None, 5000) 505000 ====================
=============================================
 Total params: 1,005,100
 Trainable params: 1,005,100
 Non-trainable params: 0
```

This is a lot of code. Let's try to break this down:

I.   First, we define the encoding layer dimensions (bottleneck layer) with `encoding_dim` and the number of features (which is the same as the number of columns in the dataset) with `numb_of_features`.

II. Next, we define the autoencoder in full:

i.   For this, first, we assign a new instance of the `Sequential` model type to a variable called `autoencoder`.

ii.   Next, we add a densely connected layer that has `encoded_dim` or the number of neurons corresponding to the learned representation of the model, a `'reLu'` activation function, and an `input_shape` parameter. Unlike the `Functional` API that you have learned in *Chapter 4, Deep Learning for Genomics*, there is no need to define the input layer here as the input layer is defined under the hood.

iii. Finally, we add another densely connected layer that accepts the learned representations from the `encoding_dim` parameter, converts them back into `numb_of_features`, and is referred to as a `decoder` block. This has `sigmoid` as an activation function.

5.   **Compile the autoencoder to prepare for training**: We should compile the model first before we can train the model. Keras' Sequential API makes this very easy for us. With a single line of code, you will be able to compile the model, as shown here:

```
autoencoder.compile(optimizer="adam", loss='mse')
```

Here, we have taken the autoencoder model and added an *Adam* optimizer, which is a very popular algorithm for optimizing models, and added MSE as the loss function for us to understand the level of reconstruction that happened during model training.

6.   **Fit the model**: Now, we are ready to fit and train the model with the compiled model from the previous step. Again, this is very easy to do in Keras:

```
hist = autoencoder.fit(np.array(pancan_rnaseq_df_
train), np.array(pancan_rnaseq_df_train), shuffle=True,
epochs=10, batch_size=50, validation_split=0.2)
```

Here, we feed `pancan_ranseq_df_train` both as features and targets. The other parameters included here are `epochs` and `batch_size`, which are hyperparameters that you can tune (you will learn more about this in *Chapter 11, Model Deployment and Monitoring*). For this example, we will use 10 epochs and a batch size of 50. Finally, to prevent overfitting the model during training, we ask it to use 20% of training data as validation data for cross-validation purposes.

7.   **Visualize training performance**: Let's quickly visualize how the model performed after training. For this, we will compare the training loss with the validation loss:

```
history_df = pd.DataFrame(hist.history)
ax = history_df.plot()
ax.set_xlabel('Epochs')
ax.set_ylabel('Reconstruction Loss')
```

This will generate a reconstruction loss like the one shown next. `loss` here indicates the training loss and `val_loss` indicates the validation loss. As expected, with more epochs, the training loss goes down, and similarly, the validation loss is shown to be going down, which indicates the model is not overfitting:

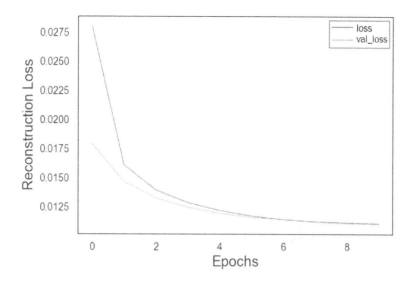

Figure 7.11 – Reconstruction loss visualized

> **Note**
>
> Please note that we only have 10 epochs for this training, and with more epochs and other hyperparameter optimization, this model loss will be even lower. But this is a good starting point.

8. **Model predictions and calculations of reconstruction loss**: Before we accept this model, we should check to see how well the model has done concerning reconstruction. You can do this in Keras very simply by running the following commands:

```
reconstruction = autoencoder.predict([input_sample])
reconstruction
array([[0.7896689 , 0.10001165, 0.5899597 , ...,
  0.08822288, 0.32368636, 0.51022756]], dtype=float32)
```

The output here indicates the difference in the gene expression values between the original input and the reconstructed output. This is just one example, but you can imagine predicting the reconstruction loss for the whole dataset and looking at the average reconstruction loss for all the genes in the DataFrame. You can also look at the most and least activated nodes from the predictions and, most importantly, plot the distribution for all 100 encoded features.

Summary

The application of unsupervised DL through learned representation is becoming extremely popular in genomics because of the large-scale datasets produced using NGS technologies. Autoencoders are being routinely used by researchers because of their promise and success across many genomics applications. Autoencoders learn by the reduced representation of the data through compression and reconstruction. During the process, they learn the key features of the data and identify the data structure automatically from examples rather than through handcrafting by humans. Diverse types of autoencoders exist to ensure that the reduced representation of the data identifies the key attributes of the original data. Autoencoders have several applications in genomics, mainly in gene expression analysis. With tools such as ADAGE, autoencoders are helping genomics datasets with no labels get biological insights from that data. We started the chapter by understanding what is unsupervised learning and the different algorithms that are currently available for unsupervised learning. We learned that Autoencoders is how they learn the key features of the data and identify the data structure automatically from examples rather than through handcrafting by humans. We then looked at the unique architecture of Autoencoders that enable us to learn these reduced representations and help address the challenging problems in genomics such as gene expression, denoising, and so on. Autoencoders have different types depending on the applications and we have a few examples such as denoising autoencoders and sparse autoencoders. Toward the end of the chapter, we took a publicly available dataset and build an Autoencoder model to predict gene expression with few lines of code.

One of the main limitations of autoencoders is that if there is no structure in the data—for example, if each data point is independent of the other in the input data—autoencoders will fail. If there is a correlation in the input features, then autoencoders will work. This is because autoencoders learn how to compress the data based on key attributes—for example, the correlation of the input features identified during the training process. Another limitation is that autoencoders can only work on the data that they have been trained on. That means they will only be able to compress the data similar to the data that they have been trained on. Despite these limitations, autoencoders have been widely used in many genomics applications, ranging from denoising scRNA-seq, time-series gene expression, identification of biological patterns, dimensionality reduction for data visualization, and so on.

We will look at the GAN generative model NN architecture in the next chapter.

8

GANs for Improving Models in Genomics

One of the significant developments in the field of **Deep learning (DL)** has been the introduction of new generative models. The most popular generative models are **Generative Adversarial Networks (GANs)**, **Variational Autoencoders (VAE)**, **deep autoregressive models**, **style transfer**, and so on. We learned about what VAEs are in the previous chapter. GANs have become a hot topic in the DL research community in the last few years. They were introduced by Ian Goodfellow in 2014 and are considered one of the most interesting ideas of the last 10 years by Yann LeCun, who is considered the father of modern DL. A GAN, as the name suggests, is a type of generative model that is trained in an adversarial setting to learn data distribution that is closer to the real world, thereby generating synthetic data inexpensively. GANs have revolutionized many domains such as **natural language processing (NLP)**, **computer vision (CV)**, and, most recently, **genomics** because of their ability to learn the data distribution and recreate artificial datasets closer to real-world data that can be used for data exploration, running tools, generating queries, test hypotheses, and so on. Since their advent, several variants of GANs were introduced, resulting in improved performance in image generation, text generation, voice synthesis, creation of artificial genomes, and so on.

Genomics data is the most complex compared to other types of data out there. Despite its complexity, the enormous potential of genomics datasets for both basic and applied research has fueled interest among genomic researchers and scientists. One of the challenge with genomics in healthcare is data privacy. GANs, because of their ability to generate artificial data derived from real-world datasets, can address data privacy. In addition, GANs can aid collaboration among researchers and the testing of ideas through open data access. GANs have several applications in genomics in addition to improving models, such as the automated design of probe sequences to perform binding assays for proteins and DNA, optimization of genomic sequences for the cell to produce a favorable chemical product, and so on. This chapter introduces GANs and how they can be used to improve models trained on genomics data. By the end of the chapter, you will know what GANs are, understand the challenges of working with genomics datasets, how GANs can improve models trained on genomics datasets, and finally, the applications of GANs in genomics.

As such, here is a list of topics that will be introduced in this chapter:

- What are GANs?

- Challenges in working with genomics datasets

- How can GANs help improve models?

- Practical applications of GANs in genomics

What are GANs?

Before we discuss GANs, you should know how generative models work. But before that, it would be advisable to understand how generative models are different from discriminative models.

Differences between Discriminative and Generative models

DL models can be broadly divided into discriminative models and generative models. Simply put, discriminative models focus on generating predictions of labels from the features mainly used for **supervised learning (SL)**, and generative models focus on explaining how the data is generated and are used for **unsupervised learning (UL)**. Let's go into this a little deeper to understand the differences.

Discriminative models try to find the relationships between X, such as features, and y, such as targets. For example, if you are trying to predict the cancer type from genomic variations (**single nucleotide polymorphisms**, or **SNPs**), the X here indicates the features of those data instances such as the number of variations, type of variation, and so on, and the y here indicates cancer type. So, the discriminative model if expressed mathematically refers to $p(y|x)$, which is the probability of a sample instance belonging to a particular cancer type y based on some feature X. As shown in *Figure 8.1*, during training, discriminative models learn the boundaries between classes or labels in the dataset:

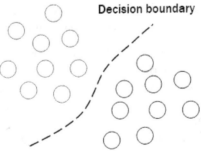

Figure 8.1 – How do discriminative models work?

The predictions from discriminative models can refer to either predicting a continuous value (regression method) or a distinct class value (classification method) as shown in the *Figure 8.1*. **Neural network (NN)** architectures such as **feed-forward NNs (FNNs)**, **convolutional NNs (CNNs)**, and **recurrent NNs (RNNs)** are examples of discriminative models.

While discriminative models learn the conditional probability $p(y|x)$, generative models learn the joint probability $p(x,y)$ of the input data x and the label y, and make predictions using the **Bayes algorithm** to calculate $p(y|x)$. Going by the previous example, a generative model knows how the data was generated, and based on the general assumption of the data, it classifies the samples into different cancer types as shown as follows:

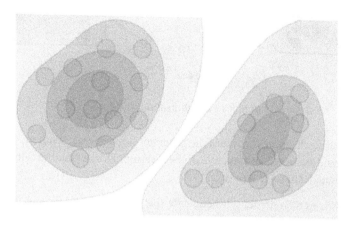

Figure 8.2 – How do generative models work?

A popular application of a generative model in genomics is representing sequences as generative **hidden Markov models (HMMs)**. Generative models are hard to train, and so they didn't take off in comparison to discriminative models; however, they have played a key role in the history of **machine learning (ML)**.

Now that you have got some background on what generative models are, let's now dive into understanding GANs for genomics in the next few sections. But before that, let's get some intuition about GANs first.

Intuition about GANs

GANs are a type of deep generative model that uses two competing (and cooperating) NN models referred to as a **generator** and a **discriminator**. As the name sounds, the generator generate the data distribution, and the discriminator acts as a critic and evaluates the generated data for authenticity. In other words, the discriminator decides whether the generated instance of data that it reviews belongs to the original data or not. Initially, we pass noisy random input data to the first NN (generator), which then tries to generate synthetic data that is statistically similar to the input data. On the other hand, the second NN (discriminator) is trained to differentiate between real and fake (synthetic) data (*Figure 8.3*):

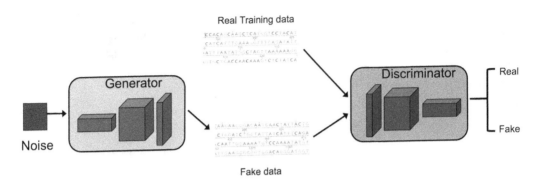

Figure 8.3 – Intuition about GANs

As shown in the preceding diagram, the role of the generator is to generate synthetic or fake data from input data from the random noise, and the role of the discriminator, which is a fully connected NN, is to classify the generated data as real or fake.

A generator in GANs tries to generate synthetic data that resembles real data. A discriminator, on the other hand, is trained to estimate the probability distribution that the given sample comes from real data rather than the one provided by the generator (fake data or synthetic data). This way, both NNs pit against each other in an adversarial way: the generator tries to overcome the discriminator, while the discriminator tries to get better at identifying whether the generated samples are real or fake. Quite regularly, the generator tries to fool the discriminator into thinking the output data is real and asks the discriminator to label it as real data, but the discriminator naturally classifies this data as fake, as expected. The game stops when the discriminator fails to differentiate whether the output is coming from real data or the generator.

How do GANs work?

Now that you have an intuition about GANs, let's understand how they work. The main components of GANs include the following:

- The input random noisy data
- The real or original data
- The generator NN
- The discriminator NN

The following diagram provides a visual representation of how GANs work:

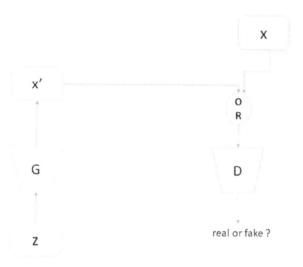

Figure 8.4 – How GANs work?

Let's get an intuition of how GANs work with the components listed in *Figure 8.4*:

- The generator G transforms a latent variable z (noisy data) to generate data x'. The structure of the generator G shown in *Figure 8.4* can be arbitrary. It can consist of an FNN or a CNN, or any other NN, and the only requirement is that the shape of the input matches the shape of the output data.

- Next, the discriminator D input is either real *data* x that comes from measured observations or fake data from generator x', and it outputs real number $D(x)$ or $D(x')$. To differentiate between the two types of data, the real data is labeled as 1.0, indicating 100% confidence in real data, and the fake data as 0.0, indicating 0% confidence in being real. The role of the discriminator is to identify real data from fake data. That means it's a binary classifier that distinguishes between real and fake datasets. As with the generator, you can choose the NN architecture of the discriminator if the dimensions of the input and output are the same.

- The process of training GANs is considered a *zero-sum minimax* game where one network's failure is the other network's gain. The discriminator aims to reduce the error between real and fake data, and the generator aims to increase the probability of the discriminator making a mistake. The loss of the discriminator reflects the accuracy of its predictions, and the loss of the generator is the inverse of the loss of the discriminator. The result is for the generator to produce samples closer to real-world data to fool the discriminator. Training generator and discriminator NNs at the same time is generally not stable.

Training of the discriminator

To train the discriminator NN, we label real data from the training set as 1 and fake data from the generator as 0.

The following diagram depicts the training process:

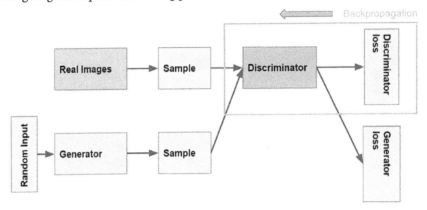

Figure 8.5 – Training of discriminator

The following is shown in *Figure 8.5*, for each batch of training, the following applies:

- The discriminator receives two sources of samples—real and fake data. The real data consists of images of real people, sequences of real DNA data, and so on.

- The discriminator uses these as training data. Real data is considered positive data and is provided to the discriminator during the training process. Fake data generated by the generator is treated as negative data during training.

- When the discriminator trains, the generator does not train and its parameters such as weights and biases are kept frozen. The role of the generator during that state is to provide the discriminator with the samples it needs to train.

- The discriminator is connected to two different loss functions—the discriminator's loss and the generator's loss. However, during discriminator training, it ignores the generator's loss. Based on the discriminator's loss it penalizes for wrongly classifying the real data as fake and vice versa, and the parameters are updated from the discriminator loss using backpropagation.

Training of the generator

Once the parameters of discriminator are updated so that the output is closer to the real-world data, we train the generator. The different steps in the training of the generator involve the following:

1. The generator produces output from random input noise samples.

2. Then, it gets feedback from the discriminator as real versus fake.

3. Then, it calculates loss from the discriminator classification.

4. Then, it uses backpropagation from the output to the discriminator to the generator to obtain gradients.

5. Finally, it uses these gradients to adjust the parameters of the generator's weight while keeping the discriminator's weights unchanged.

Let's go into the details of each of the preceding steps for the training of the generator.

You can see a visual overview of the training here:

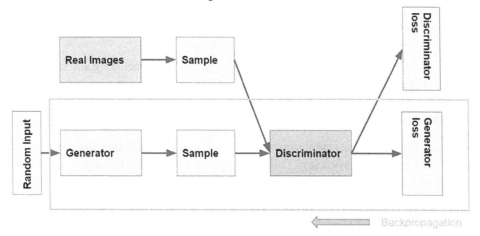

Figure 8.6 – Training of generator

The following is shown in *Figure 8.6*:

1. The generator produces output from random input noise samples and is then passed to the discriminator to compare it with the real output and report output loss.

2. The generator gets feedback from the discriminator as to whether it is real or fake and tries to learn and improve model predictions over time.

3. The generator has a loss like a discriminator, and it penalizes the generator for producing a sample that the discriminator classifies as fake. As before, the parameters of the generator are adjusted using backpropagation, and during this process, the discriminator's parameters are kept frozen.

4. When the generator does a good job to generate good data and fool the discriminator, the output probability should be close to 1.

With this background, let's understand how GANs can help genomics but before that let's look at the current challenges working with genomic datasets.

Challenges working with genomics datasets

Genomics is the study of the genetic constitution of a whole organism, which are instructions for an organism to build and grow. It is now routinely possible to sequence a whole genome of organisms, thanks to **next-generation sequencing** (**NGS**) technologies. Despite easy access to genome sequencing technology, the primary challenge is the availability of these genomic datasets at scale because of technical limitations, cost, difficulty collecting more data, and so on. It is well known in the DL community that in general, the more data that DL can have access to, the more accurate the predictions are.

Not having enough data restricts the utility of the available data and limits building highly accurate DL models with it. Here are some of the problems arising from small data:

1. Small data poses problems with model training and the use of trained models in real-world applications because it is prone to overfitting problems.

2. Small data is also often confounded by problems such as class imbalance and bias, which is common in healthcare data. Class imbalance is where there is an uneven distribution of samples in one class versus another and results in models converging slower during training and poor performance on real-world datasets during model predictions.

In addition, to small data, the other big challenge for genomics remains safely sharing genomic datasets because of sensitivity and privacy issues. Our ability to safely share this data will lead to medical breakthroughs and precision medicine reality in near future.

One potential solution to address the aforementioned problems is to create synthetic datasets that can mimic real-world genomic data and can offer enhanced data privacy and enable better collaboration among genomic researchers without compromising privacy and bias. In addition, they also augment existing real-world genomic datasets. GANs can generate artificial datasets of the highly complex genomic sequences that genomic researchers routinely use. One of the main advantages of GANs is the artificial genomic datasets that have the same size and shape as the real-world dataset it was trained on, which makes GANs the preferred choice for generating synthetic data. In addition, researchers can run tools, test hypotheses, and do any other analysis on this synthetic data, similar to how they do it on real data. Before we go further, let's understand what synthetic data is.

What is synthetic data?

As you all know, real data is obtained from direct measurement and is constrained by cost, logistics, and privacy concerns in the case of human data. In contrast, synthetic data is artificially synthesized and is closer to the real-world data generated using generative models such as GANs. Synthetic data can overcome many of the limitations of real-world data and, according to Gartner, *by 2030 the use of synthetic data will overshadow real data in AI models* (`https://www.forbes.com/sites/robtoews/2022/06/12/synthetic-data-is-about-to-transform-artificial-intelligence/?sh=fbe727c75238`). This is mainly caused by several factors such as an increase in compliance costs, regulatory restrictions because of the **Health Insurance Portability and Accountability Act (HIPAA)**, the **General Data Protection Regulation (GDPR)**, the **California Consumer Privacy Act (CCPA)**, ever-increasing cybersecurity attacks, data breaches, privacy concerns, increased costs for manual annotation of data for DL. Synthetic data plays a key role in the advancement of DL, especially in genomics, because of limitations of the data shareability and a lack of relevant data, which is common in many genomic applications to train a model effectively.

Synthetic data can augment real-world training data and address these limitations. There are several use cases of synthetic data, especially those instances where positive cases are very rare compared to negative cases—for example, cancer data where the number of samples from cancer is far fewer than the number of non-cancer patients.

Synthetic data for genomics

Synthetic data can augment limited genomic datasets that are available for many organisms. This way, we can improve the DL models to produce better and more accurate results and also prevent algorithm biases and overfitting. For example, it is widely known that the public datasets available out there are unbalanced in terms of gender, race, geography, and so on. Because of this imbalanced nature of datasets, the algorithm is biased toward the majority class (the class that has more data points), and classification accuracy suffers because of this. If we were to use this model, it would work great for the majority class but would have poor performance for the minority class (the class that has low data points), although its performance for the minority class is important compared to the majority class.

There are techniques such as oversampling and adding a penalty term to the cost function for the wrong prediction, which we will discuss briefly next:

1. Oversampling is where instances of the minority class are duplicated. The **Synthetic Minority Oversampling TEchnique** (**SMOTE**) is a very popular oversampling technique that creates synthetic samples by interpolating between neighboring minority classes in "feature space" rather than in the "data space". It uses the **k-nearest neighbors** (**KNN**) method and creates a synthetic sample at a random point on the line between each data point and a chosen neighbor. Even though SMOTE has been widely adopted by ML practitioners and has been successfully used for unbalanced datasets, it's proven to be suboptimal for the generation of synthetic data for high-dimensional datasets such as NLP, CV, genomics, and so on.

2. Cost-sensitive learning, where we modify the cost function to penalize the misclassification of the minority class. However, it also suffers from the same issues as oversampling method.

3. Synthetic data generation using GANs. Generative models such as GANs, because of their ability to learn data distribution from the trained data, can create synthetic datasets with the same size and distribution as the original data that they have been trained on and thereby can augment high-dimensional datasets such as images, audio, videos and, genomic datasets and address class imbalance. Several published pieces of the literature showed the promise of GANs for augmenting the trained dataset and thereby improving performance in the case of imbalanced high-dimensional class-imbalance problems. The three main ways synthetic datasets can help researchers, developers, scientists, and enterprises are summarized as follows:

    -   Making data accessible and shareable, thereby allowing for faster and safer collaboration on the data and arriving at interesting and innovative findings sooner

    -   Generating more samples from the limited datasets can help models generalize well on unseen data

- Reducing bias in training datasets, which in turn can help build representative and highly accurate models thereby improving models.

You now have some background about what GANs are, how they work, and how GANs can help address some of the current challenges in genomics. With this background, let's spend some understanding how GANs can help improve models.

How can GANs help improve models?

DL requires a lot of data to mine insights and make an informed decision. The success of DL to generalize well is mainly attributed to the training of NN architectures on large amounts of data. However, it is not always possible to acquire more data because of several reasons, as explained earlier. What if we can generate synthetic data that is modeled around real-world data so that we can augment the limited datasets and improve our model predictions? Synthetic data has a multitude of use cases in DL because of the infinite variations of synthetic data that can be produced. DL is the primary beneficiary of synthetic data, and research shows that enhancing real-world data with synthetic data produced using generative models such as GANs can significantly improve model fitness and thereby result in better predictions. GANs can help improve models directly and indirectly through the generation of synthetic data, which can make sensitive data accessible to help researchers understand the features in the data that best explain the problem. It can also be used to augment limited datasets—which are quite common in genomics because of technical limitations, costs, and feasibility—to balance data to reduce bias and improve predictions for models, which is the most important component of what GANs do.

Before we understand how GANs can help improve models, let's refresh our understanding of how synthetic data is generated. Briefly, synthetic data is generated from real-world datasets by using generative models such as GANs that look at the distribution of data points in the training data and resample data from the real-world data, as shown in *Figure 8.7*:

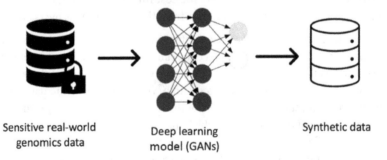

Sensitive real-world   Deep learning   Synthetic data
genomics data   model (GANs)

Figure 8.7 – Synthetic data generation process using GANs

Let's see an example of how synthetic data using GANs can be produced using a real-world case study.

Scientists often explore the relationships between the phenotype (physical characteristics of an organism height, weight, and so on) and genotype (genetic differences that exist at specific locations such as SNPs and indels) to identify genes controlling a particular disease, a process termed **genome-**

wide association studies (GWAS). One such widely-used repositories to study populations is UK Biobank, which hosts the phenotypic and genotypic data from a large number of individuals and has more than 400 billion data points. For this example, we will take data set of 1,200 mice that contains 68 phenotypes and nearly 100,000 SNPs.

Let's now use GANs to create synthetic versions of this real-world dataset one for phenotypes and one for genotypes from the aforementioned data. Once we generate synthetic datasets, can then compare the statistical results of the real-world and synthetic datasets to find out if synthetic datasets are any good compared to real-world datasets. Here are brief the steps:

1. The first step is to build a phenotype training set where we create and format real-world training data for each phenotype batch that will be generated.

2. Next, from the phenotype training data, train synthetic model to generate synthetic phenotypes matching their size and shape.

3. Similar to building a phenotype training set, we build a genotype training set where we format and build a training set for genomic data along with synthesized phenotype data.

4. And finally, we train a synthetic model on real-world genotypic data and ensure that the synthetic phenotypes line up with our newly created synthetic genotypes.

5. Once a synthetic version of the data is generated, we can do an initial analysis of this data. We can compare the accuracy of the synthetic genotype and phenotype data to the real-world training sets through a correlation matrix. We can also do **principal component analysis (PCA)** between the two datasets to examine how effectively the synthetic model learned the structure and distribution of the real-world data.

6. Once the synthetic data is validated by comparing it with the real-world data through correlations and PCA, we can now leverage this synthetic data for genomics-related use cases.

We have just seen a use case of how GANs can help improve models by creating synthetic datasets. Similarly, there are several other applications of GANs for genomics that we will learn about in the last section of this chapter.

Practical applications of GANs in genomics

GANs have found a lot of applications in several domains such as NLP, CV, and genomics because of their ability to produce synthetic data samples to augment the real world and help improve models' fitness. State-of-the-art synthetic models such as GANs can produce an artificial version of high-dimensional and complex genomic datasets with high accuracy, scale, and privacy. The artificial datasets can be shared among researchers and enable future genomics research and safe, private data sharing between researchers, health care providers, and the industry. As discussed briefly in the introduction, there are several use cases of GANs in genomics such as the automatic design of probe sequences for binding assays, optimization of genomic sequences, creation of synthetic genomes, and so on.

We will now see some examples of how GANs are applied to genomics and solve some real-world problems in the following section.

Analysis of ScRNA-Seq data

Single-cell RNA-Seq (scRNA-Seq) technologies have enabled gene expression profiling at a single-cell resolution level because of the advances in Single-cell and NGS technologies. ScRNA-Seq gene profiles enable the understanding of the function of genes at a single-cell resolution level like never before. There is a lot of ScRNA-Seq data that is currently available, to extract meaningful biological insights. Despite the availability of large amounts of data, integration and analysis of **scRNA-Seq** data is computationally challenging because of biological and technical noise coming from different laboratories and between different batches within the same experiment. Current methods such as dimensionality reduction, clustering, and so on can help find the structure in the data when the data is devoid of any noise, but they cannot help integrate the data coming from diverse laboratories and experimental protocols.

GANs can be applied to a few types of scRNA-Seq data originating from different labs and different experimental protocols and can generate realistic scRNA-Seq data that spans the full diversity of all the cell types. This is a powerful framework for the analysis of scRNA-Seq data by integrating multiple datasets irrespective of the origin and experimental protocols. Let's understand how this is done now.

Have a look at the following diagram:

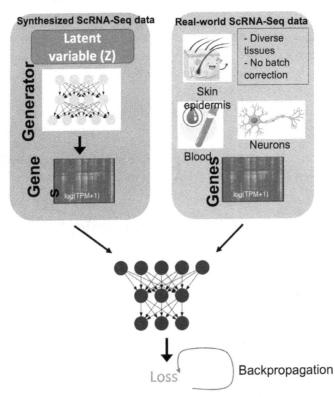

Figure 8.8 – An overview of single-cell RNA-Seq analysis using GANs

In the preceding diagram, GANs with the two NN models—the generator and discriminator—concurrently train and compete. The role of the generator, as learned before, is to transform data from the latent variable or random noise into a single-cell gene expression profile. Next, the discriminator tries to evaluate whether the generated data is closer to the real-world data or generated. Only the discriminator sees the real scRNA-Seq data, which is not corrected for batch effect and technical variation, while the generator tries to improve its synthetic data through its interaction with the discriminator. Once the GAN is optimized by adjusting the generator and discriminator parameters using backpropagation, we can extract the biological insights from the generator and discriminator NN models such as gene association networks, gene expression ranges, dimensionality reduction, and so on.

Generation of DNA

The generation of DNA is the most common application of GANs in genomics. Deep generative models such as GANs can be used to model DNA sequences. DNA sequence data is unique and can be considered a hybrid between NLP and CV. DNA is the most natural language ever, consisting of the 4 nucleotides - A, G, C, and T organized on a gene with a hierarchical structure such as introns exons. As with a CV model, they have regularly repeating patterns (motifs). GANs can be used to create DNA sequences that have desired characteristics.

The framework shown in *Figure 8.9* is used to create a DNA sequence:

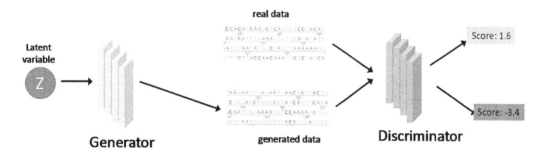

Figure 8.9 – Basic schematic of GAN for generation of DNA

This model consists of the same two components—generator and discriminator—and they are both used for training this architecture.

- As we have learned before, a discriminator is trained using real data and fake data. Here, the real data is taken from the real-world dataset, and the fake data has obtained from the generator by sampling the random noise data (latent variable z).

- During the training of the generator, synthetic data is generated by sampling the random noise data (latent variable z). and sent to the discriminator to be scored to maximize scores for the synthetic data.

- The discriminator tries to maximize the score of real-world data taken from the real-world dataset and minimize the scores of fake data generated by the generator.

- Both models are trained using standard gradient descent algorithms. Finally, after training, we can use the generator for the creation of synthetic data and, in this case, DNA sequence data.

Using GANs for augmenting population-scale genomics data

As we saw previously, even though NGS has made it possible to generate large-scale data through sequencing genomes quickly and cheaply, it is not always possible to generate data—for example, in the case of rare diseases where samples are limited, and so the generation of data from those samples is hard. In addition to this, data privacy is an issue when accessing human data for research and other purposes. If we decide to build models on the available data, which is small and unbalanced, then the models will be biased and conclusions will be error-prone. Previously, we saw an example of how GANs can help address this problem. Unfortunately, GANs are super hard to build in Keras, so we will not do a hands-on analysis. Readers are encouraged to check out the following link for detailed instructions on how to build GANs using Keras from a similar dataset: `https://github.com/shilab/PG-cGAN`.

Summary

DL algorithms have seen a major upgrade recently with the development of generative models such as VAEs and GANs, which contributed significantly to the creation of synthetic datasets. With this development, the fields of CV, NLP, and genomics have profited immensely. In the last chapter on Unsupervised learning using Autoencoders, you were introduced to VAE and in this chapter, you were introduced to GANs and how they can be used to address some of the limitations of genomics data and, improve DL models. First, we looked at the differences between discriminative and generative models, and then next we understood the key components of GANS which are the generator and discriminator, how they are trained and constantly pit against each other in an adversarial way to generate synthetic data as close as possible to real-world data.

Because of GANs ability to generate synthetic data and DL's requirement for a large amount of data, we see how GANs are used for improving models—specifically, DL models. Then we had a case study where we understood how GANs can help create synthetic datasets to augment the phenotype and genotype real-world datasets and perform accurate GWAS analysis. Finally, we saw some of the important applications of GANs in genomics such as the analysis of heterogenous scRNA-Seq data and the creation of artificial DNA sequences. We also realized that these could not have been possible if it were not for GANs, and it's only the start of what GANs can do for genomics.

In the next chapter, we will look at how the different **deep NN (DNN)** algorithms such as FNNs, CNNs, RNNs, autoencoders, and GANs that we have learned so far can be operationalized in a production environment.

Part 3 –
Operationalizing models

This part will describe operationalizing models, which includes training, tuning, validating, deploying, and monitoring models. In addition, this part will describe model interpretability along with the pitfalls, challenges, and best practices for machine learning and deep learning technologies for genomic applications.

This section comprises the following chapters:

Chapter 9, Building and Tuning Deep Learning Models

Chapter 10, Model Interpretability in Genomics

Chapter 11, Model Deployment and Monitoring

Chapter 12, Challenges, Pitfalls, and Best Practices for Deep Learning in Genomics

9

Building and Tuning Deep Learning Models

Deep learning (DL) algorithmic solutions are currently being leveraged in several biological and life sciences domains to address some of the most challenging problems in healthcare, medicine, agriculture, genomics, and so on. Among these disciplines, genomics poses extreme challenges to DL because of how complex the field of genomics is, which goes way beyond the knowledge of how to interpret genomes. Thankfully, a lot of genomics research with DL has led to the design of sophisticated **deep neural networks (DNN)** architectures that are suited to genomic tasks. This intersection of DL with genomics proved very successful, leading to the application of DL to several genomic applications in regulatory genomics, functional genomics, structural genomics, and so on. Furthermore, it allowed the genomics research community to gain a global perspective of the human genome, which is paving the way for the goal of genomic medicine in near future.

So far, we have covered some basic and advanced concepts of DL and their application in genomics without going into detail about how to build those models. Successful application of DL algorithms in genomics requires more than just the basic concepts of which algorithms exist and what they do. It is important to know how to practically apply these algorithms in genomics. In this chapter, you will take those concepts and build DL models to address a few genomics problems. By the end of this chapter, you will have learned how to process genomics data, extract and select features from genomics datasets, train a DL model, tune hyperparameters, and perform a model evaluation. As such, here is the list of topics in this chapter:

- DL life cycle
- Data processing
- Developing models
- Tuning the models
- Model evaluation
- Use case—Predicting the binding site location of the JunD TF

Technical requirements

Google Colab notebooks: Because we will be training a **convolutional NN (CNN)** on slightly larger datasets than our laptop CPU could handle, we will develop our models in Google Colab. Colab is a free-to-use Jupyter notebook-kind of framework that provides a free **Graphical Processing Unit (GPU)** to train models. We will learn the basics of Google Colab in the following section, but readers can find more details about Colab here: `https://towardsdatascience.com/getting-started-with-google-colab-f2fff97f594c`. The key to building models in Google Colab is selecting a GPU so that you can build your model faster. You just go to the **Runtime** drop-down menu, select **Change Runtime type**, and then select **GPU** in the **Hardware accelerator** drop-down menu, as shown here:

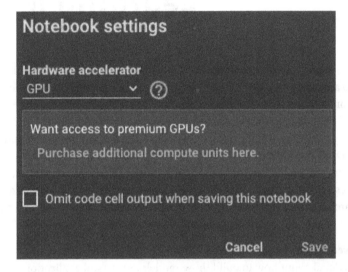

Figure 9.1 – Google Colab notebook settings

Once a GPU has been selected, you can run the commands shown in the hands-on section of this chapter.

DL life cycle

The DL life cycle is quite complex as it involves several phases, starting from identifying a business goal to model monitoring (see *Figure 9.2*). Each phase is driven by the key decision of what, why, and how that will affect the entire DL life cycle. Previously, we have looked at the basic concepts of leveraging DL for genomic applications, but to leverage DL to genomics requires one to understand each of these phases in the DL life cycle clearly. With that goal in mind, let's go into the details of each of these phases in the following section:

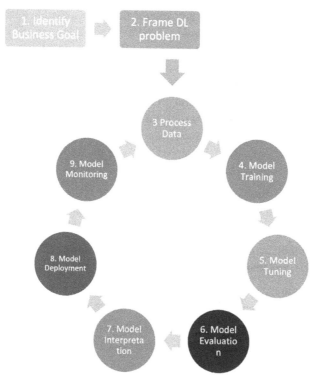

Figure 9.2 – Different phases of a typical DL life cycle

We will go through the steps given in the previous screenshot:

1. **Identify business goal:** The first phase in the DL life cycle is identifying a business goal. Every DL project starts with a business problem, scope and risks. This is simply coming up with a clear idea of the business problem to be solved any kind of risk that the business can run into, how will we know if we have met our objective, what is the simplest of solutions that can help, A key watchout at this step is choosing a business problem to solve where there is a solution in place. and the business value that the solution brings to the company. Once you have a clear idea of the business problem, next is the determination of success criteria or **key performance indicators** (**KPIs**) to evaluate the company's ability toward that target. Please keep in mind that the target should be attainable and provide a clear path to production. The evaluation should help answer if DL is the appropriate technology for solving the business problem, along with other options that are available to achieve that goal. For example, there might be other simpler methods such as statistics or ML models that can solve the same problem. The key to the success of building a successful DL model is the availability of data, hence a thorough investigation of the available data sources and accessibility should be the topmost priority. This should be done after a thorough analysis of the business goal. This is call feasibility analysis where we assess the feasibility by exploring the raw data. This is where most of the time is spent on real-world projects..

2. **Frame a DL problem**: The next phase is to convert the business goal to a DL problem. In other words how do we want to solve the problem and then conveying the solution to the business users. . There are two key components of this phase—what to predict and what are business and model performance metrics and how does it look like which is extremely important to understand.. Let's suppose a biotech company wants to identify drugs that can treat rare diseases and bring maximum profitability. Reaching this business goal partly depends on the types of drugs that the company wants to bring to the market and the market demand for these drugs. Once business stakeholders are aligned on the problem and the final outcome (in this example, drug-response predictions), the next step is to explore the trade-off between state-of-the-art methods vs DL based approaches.

3. **Process data**: Post-DL problem framing, the next phase is the data processing phase, which includes data collection and data wrangling The data wrangling phase in turn includes data preprocessing and feature engineering.

4. **Model development**: Model development includes a training phase, a tuning phase, and evaluating models, and this is the crux of DL.

5. **Model interpretation**: Model interpretation is important for understanding how and why a particular model worked, and this is especially important for genomics.

6. **Model deployment**: Model deployment is when you put the evaluated model in production for final consumption and prediction purposes, either through a web app or serving it with an API.

7. **Model monitoring**: Finally, model monitoring is useful for identifying indicators of data and concept drift.

The DL life cycle is a highly iterative process with frequent loops of data collection, data preprocessing, model creation, model tuning, model retraining, and model testing. The overall process can be summarized as follows:

1. Use the collected data to process the dataset, and create the model.

2. Tune the model or refine the dataset depending on the model evaluation results.

3. Create a new model and compare it with the previous models.

4. Create a final production model from the most recent dataset.

Generally, *steps 2-4* from above are repeated until the model generalizes very well on unseen data, and then the model is deployed. The DL life cycle is iterative post-deployment too where you will continue to monitor the model in the production environment for any model performance issues and continue to find ways to improve the model.

In this chapter, we will cover the DL life cycle from data processing to model development. In the next three chapters, we will cover the rest of the DL life-cycle phases (model interpretation, model deployment, and model monitoring). It is beyond the scope of this chapter to cover the first 2 phases (identify business goal and frame a DL problem).

Data processing

In the DL life cycle, the data (inputs and outputs) serves important functions such as defining the goal of the problem, training the algorithm, evaluating the performance of the trained model, and building baselines for model monitoring and so this is considered as the most important phase of the DL life cycle. The data processing phase can be subdivided into data collection and data wrangling, which in turn divides into data processing and feature engineering, as depicted here:

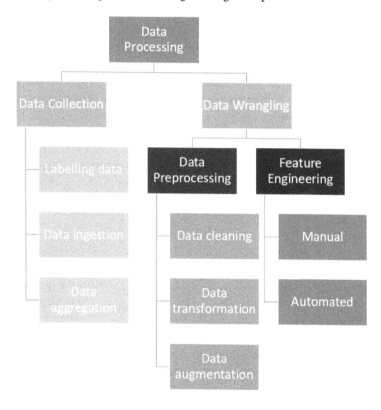

Figure 9.3 – Different subphases of data preprocessing

As shown here, the data collection phase mainly includes identifying data resources and the accessibility of data. The data wrangling phase includes data preprocessing and feature engineering. Let's discuss each of the phases in detail in the following section.

Data collection

Data collection is technically the first step of the DL life cycle. Without data, there is no model. Data collection generally refers to identifying a data source for building models for a particular problem in genomics. There are some key questions related to data collection, such as what type of data to collect, how much data to collect, and so on. All of this depends on the business problem as well the biological variability of the available data. Unlike other domains, genomic data collection is a very complex process. Genomics data can be collected from publicly available datasets and open-source databases. Some popular genomic repositories include NCBI GEO, SRA, TCGA, and so on. Many times, it requires the experimental generation of data, which is very common in genomics. Once you identify the type and source of data collection, then evaluate various ways of collecting the data. Another key to data collection is the type of data—for example, labeled data or unlabeled data. Most genomics data doesn't come with labels, and hence efforts need to be put into either generating labels manually or in an automated fashion or using unsupervised methods such as clustering or **principal component analysis** (**PCA**). Once the data source is identified, it is recommended to use popular data technologies such as classical **Structured Query Language** (**SQL**) databases or advanced data lakes and data warehouses to aggregate all the collected data. **Extract, transform, and load** (**ETL**) technologies such as Azure Data Factory (`https://azure.microsoft.com/en-us/products/data-factory/`)or Fivetran (`https://www.fivetran.com/`)can help aggregate and orchestrate data ingestion and transformation across several sources automatically. Data collection normally takes the most amount of time. Even if you have enough data to build the model, chances are that the data requires some preparation. Let's talk about the data wrangling phase next.

Data wrangling

Data is the center of DL and is the means of creating DL models. The DL models are as good as the input data, and so data quality is a critical component of building highly optimized and generalizable models. So, the key to effective model building is determining if you have the right data and quality data. If the answer is no to either of those, then the first problem to solve is how to get the data for model building. Note that you can get more value by iterating on the data rather than the model. One of the key steps during the data wrangling stage is understanding the data through **exploratory data analysis** (**EDA**), which uses statistics and visualization methods for finding the relationships between different features, samples, and other variables. In the case of genomics, EDA can help with sample-to-sample variability based on experimental design, outlier detection, and so on. The data wrangling step can be subdivided into data preprocessing and feature engineering (*Figure 9.3*).

Data preprocessing

This step refers to the processing of raw data into a format that is suitable for either EDA or training models. The data preprocessing steps include the following:

- **Data cleaning**: Data cleaning refers to those preprocessing steps that are used to remove outliers and duplicates, filter inaccurate or irrelevant data, correct missing values, and so on. This step is necessary to remove data points that have unreliable measurements so that they don't affect downstream processes.

- **Data transformation**: Scaling the data is very important as it ensures all the features are in the same range. This will help during the modeling steps as many DL algorithms expect the data to be normalized. There are two main methods available for data transformation—normalization and standardization. Normalization ensures the data values are between 0 and 1, whereas standardization will have a mean of 0 and a standard deviation of 1.

- **Data augmentation**: The creation of artificial or synthetic data from the original data constitutes data augmentation and is critical for unbalanced datasets. Data augmentation prevents overfitting.

Feature engineering

Feature engineering is a process of selecting and engineering features of training models. A typical feature engineering process has multiple steps such as feature selection, feature transformation, feature creation, and feature extraction. However, in the case of DL, feature engineering is very minimal and mainly done in an automated way through learned features. The most exciting thing about DL is its promise that we no longer have to create hand-crafted features because features are automatically learned and extracted by the algorithms. There are instances where one can employ hand-crafted features instead of learned features. Let's understand the difference between the two different types now.

Comparing learned features and engineered features

The key to building good models is a selection of features that will ultimately be part of the model development process. Hence, feature selection is key to building good models. As discussed previously, a feature can either be hand-crafted or automated. Let's go over an example to understand the difference between when one would employ hand-crafted features versus automated features.

Suppose that you want to build a model to predict the expression of a gene because that model holds a promise to provide us with solutions to cure diseases. Furthermore, it helps us understand how gene expression regulates the development, adaptation, and growth of living organisms. So, you went and collected several gene sequences from the organisms of your interest from open-source databases such as Ensembl (https://www.ensembl.org/), GEO, TCGA, and so on.. Before **deep NNs (DNNs)**, for a given gene sequence, you would first select the important features such as gene expression values (as seen in *Chapter 3, Machine Learning Methods for Genomic Applications*), nucleotide context (% of A, G, T, and C), GC content, presence or absence of a particular motif, codon frequencies, and so on.

Next, all the non-numerical features such as categorical features are converted to numerical features (encoding), and finally, these hand-crafted features are then used for model training. This workflow for feature extraction works great if the problem is simple and only has a few features to use in the model training process. With the advent of DL, the process of feature engineering has become more automated because DNNs learn to predict features directly from the input data (in this example, gene sequences) without the need for manual feature extraction. The only input to these DL models is one-hot-encoded DNA sequences, using which, the model you select (such as a CNN) will learn to extract useful features, and the whole process of feature engineering is automated. So, instead of manually extracting the sequence-based features, you let the DL model learn them for you and use them for model building. This way, it can achieve the task of predicting gene expression from a DNA sequence at very high accuracy without the need for hand-crafted features.

So far, we have looked at the data processing phase of DL. Next, we will dive into the most exciting part of the DL life cycle, which is developing models.

Developing models

Model development is my favorite part of the DL life cycle (and probably 90% of data scientists' favorite part too). The goal of model development is to build a model with minimum loss over training and validation datasets to prevent overfitting, which is done by searching for the parameters that best fit the model. A typical model development phase involves four different steps, as shown in the following diagram:

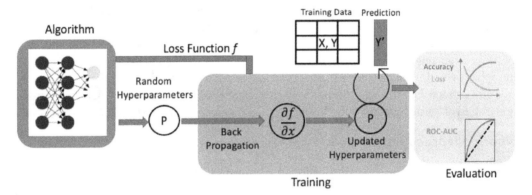

Figure 9.4 – Different steps of the DL model building phase

Let's discuss each of the four steps briefly:

1. **Selecting an appropriate algorithm**: In this step, an appropriate DL algorithm is selected based on the problem that you are solving.

2. **Model training**: Once an algorithm is selected, the DL algorithm is provided with the training data, loss function, random hyperparameters, and objective metrics to optimize using backpropagation.

3. **Model tuning**: The model initially has random hyperparameters that later get updated during this phase.

4. **Model evaluation**: The optimized model using updated hyperparameters is then evaluated on the test data using several metrics such as accuracy, ROC-AUC, AUC score, and so on.

In this section, we will go into the details of the first two, which are selecting a model algorithm and the training of the DL model, and in the next few sections, we will go into details of model tuning and model evaluation.

Selecting an appropriate algorithm

For a given DL problem, since multiple DNN architectures can solve the same problem, selection of an appropriate DNN architecture is critical. For example, for predicting a **transcription factor binding site (TFBS)** detection problem, we can either use a CNN, a **recurrent NN (RNN)**, a **bidirectional long-short-term memory (BiLSTM)**, or a **multilayer perceptron (MLP)** algorithm, and so on. Even though there are several performance metrics available for evaluating different algorithms, it is not recommended to try every DNN architecture for every problem as this requires a lot of effort and time investment. A proper understanding of how to select the right algorithm is important. In this section, we will look at some guidelines to select an appropriate algorithms for a particular problem

1. **Model interpretability**: The ability of a model to explain its prediction is loosely defined as model interpretability. We have a full chapter dedicated to this and so we will not go into the details here.

2. **Training time**: The amount of time it takes for a model to train is a key factor for algorithm selection.

3. **Memory requirements**: In general, all DNNs require a lot of memory but there are subtle differences in the memory requirements for the algorithms. If there are multiple DNNs available for a given problem, picking one with the lowest memory requirement is recommended.

4. Understanding the problem domain: This should be at the top of the list. A proper understanding of the problem domain that you are trying to solve. Fortunately for genomics, the input data is mainly sequence. However, when it comes to selecting a DNN architecture, you should think if your solution requires capturing spatial information then CNN is a good architecture, and if you are interested in capturing the temporal dynamic behavior of the data, then RNN along with its variants (LSTM, GRU, BiLSTM, and so on) are a good fit.

Model training

Once a particular DNN architecture has been selected based on the above factors, the next step is training the DNN architecture with the data.Model training is the process of training the algorithm to identify the features, and it consists of the steps covered in the following subsections.

Data partitioning

The first step toward developing a DL model is data partitioning or splitting the data into three parts: training, validation (sometimes referred to as development), and holdout (also called test) datasets using a defined percentage. For example, a typical model development process in genomics involves splitting the datasets into 70% training, 10% validation, and 20% testing, as shown here:

Figure 9.5 – Data partitioning in typical DL model development

Let's discuss what the training, validation, and holdout datasets are, as depicted in the preceding diagram.

Training dataset

The training dataset is a subset of the whole dataset that is used to train the algorithm to learn the mapping between the features and the label, in the case of **supervised learning** (**SL**). As indicated earlier, both the quantity and the quality of the training data are critical for the success of the model development. In genomics, training datasets typically constitute 70% of the whole data.

Validation dataset

The validation dataset is another subset of the whole dataset and is used to optimize the parameters and hyperparameters to develop the most accurate model that performs well on the unseen dataset or reduces overfitting and makes the best business decision possible.

Holdout dataset

The holdout dataset or test dataset is the last subset and is mainly used to evaluate the trained and validated model before it gets deployed into production. The holdout dataset should not be used for anything other than testing the performance of the final model.

One of the challenges in model development is determining the ratios of data partitioning and the method of partitioning. The ratio of data points in each partition depends on the dataset size, domain knowledge, and so on. How the partitioning of data is done is also important—for example, you don't want all the rows that have the same class in the same partition, right? If that happens, the model wouldn't have that class available for training and validation in other datasets. To address this, there are three main methods available:

- **Random partitioning**: This method randomly assigns the observation in the three different partitions (training, validation, and testing).

- **Stratified partitioning**: With this method of partitioning, there is an equal representation of classes in the three different partitions (training, validation, and testing). This is especially useful if the dataset consists of an unbalanced number of classes.

- **Cross-validation**: When you have limited data, you may not have the luxury of partitioning the data into three parts. If you do so, then your validation and holdout sets will be small, thereby not being a proper representation of your entire training data. For instance, if you have a dataset of 100 samples and you split your data into 80-10-10 splits, then you will end up with only 10 data points for both validation and holdout sets, which may be not enough to validate and evaluate your model. One solution to this problem is a technique called cross-validation. There are three different types of cross-validation, as detailed next.

K-Fold cross-validation

In this method, the data is shuffled and then split into *k* groups or folds. After splitting into *k* groups, at each iteration over each group, one group is used for validation and the other groups (*k*-1) are combined into a training set. After training, the model is validated on the first group, and the process continues *k* times by choosing a different subset to be the validation set every time. After the final iteration, the best model can be selected based on the highest score.

Stratified K-Fold cross-validation

Stratified K is like K-Fold cross-validation except that the labels or the target values are taken into consideration while splitting the data. For example, if there are two classes in the target variable, then this method ensures that each fold represents equal ratios of the two classes when compared to the training dataset. This is more accurate and produces less biased models compared to conventional K-Fold cross-validation.

Group K-Fold cross-validation

In this method, while splitting the data into various k-folds, it considers the information about groups of dependent examples such that all examples belonging to the same group are assigned to the same fold (either train or test). An example of this type includes at which chromosome the gene is located, so all samples related to a particular chromosome are on the same fold.

Tuning the models

One common problem in model development is overfitting. Overfitting happens when the model performs well on the training data but does not generalize well on unseen data. There are several reasons for overfitting, such as high model complexity, training for many epochs, too little training data, and so on. Model tuning is the processing of increasing model performance by limiting model complexity, regularization, dropout, and so on to reduce overfitting. This is generally done in DL by optimizing "hyperparameters".

Before we further discuss tuning models, let's understand the difference between parameters and hyperparameters. Parameters are inputs to the ML library or model that can be generally learned from the model. Some examples of the parameters of NNs include weights and biases. During model training, through backpropagation, the model learns those parameters, whereas hyperparameters are those parameters that cannot be learned from the model automatically and hence need to be manually adjusted during training. Hyperparameters can be thought of as dials or knobs of the ML and DL models to improve model accuracy and other metrics. Hyperparameters can be broadly divided into two types—numerical hyperparameters and categorical hyperparameters. Numerical hyperparameters are again classified into continuous (for example, learning rate, momentum, and regularization) and integer (for example, number of layers, number of batches, node size, and batch size in **stochastic gradient descent (SGD)**). Categorical hyperparameters can be further divided into a finite domain, unordered (for example, type of activation function), and Boolean (for example, use preprocessing or not).

Hyperparameter tuning

Choosing an appropriate hyperparameter is crucial for developing a highly accurate DL model. Methods to perform hyperparameter tuning can be divided into domain knowledge-based tuning and black box-based tuning. Domain knowledge-based tuning is leveraging domain knowledge to adjust hyperparameters and fine-tune the models, which does not require huge computation. You may be a statistician or scientist and you might bring domain knowledge, which lets you test some of those more from a practical perspective. Black-box tuning is the most popular type of hyperparameter tuning and is the tuning of hyperparameters using some brute-force methods such as Grid search, Random search, and Bayesian optimization.

Grid search

This method is also called parameter sweeping, where we manually define a subset of hyperparameters and then use all possible parameter combinations for the specified parameter subsets. Each combination of hyperparameters is validated using cross-validation, and then the best-performing hyperparameter combination is selected. This is considered the simplest way of hyperparameter tuning. Let's say you have two continuous hyper parameters X and Y, then you try every combination of values in those hyper parameters until you exhaust the whole space. For example, X here is the number of number of nodes (2,3,4,5..) and Y is the learning rate (0.001, 0,002) and the combination of those can be (X=2, Y=0.001), (X=3, Y=0,002) and so on. You can see an illustration of this concept here:

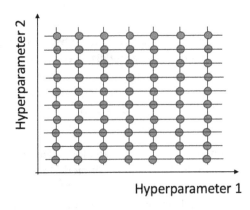

Figure 9.6 – Example of Grid search hyperparameter method

As can be seen from *Figure 9.6*, this is an exhaustive search method that spans all the combinations and is very effective in finding the best combination of parameters. However, this is very slow because checking each combination of parameters takes a lot of time and computational resources (imagine if you had more parameters to search). In addition, each combination on the grid needs K-Fold cross-validation, which requires k training steps. So, this is considered a very expensive method, but if you are looking for a simple solution for hyperparameter tuning, then Grid search is the best method.

Random search

Random search, on the other hand, is a hyperparameter method that is like the Grid search method, but instead of searching for all possible combinations, this method tests only a randomly selected subset of combinations (*Figure 9.7*). The smaller the subset, the faster the hyperparameter tuning and the less accurate the results. On the other hand, the larger the subset, the slower the hyperparameter tuning and the more accurate the results:

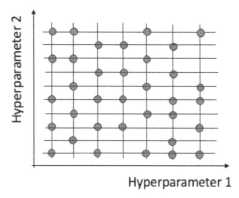

Figure 9.7 – Example of Random search hyperparameter method

This method is useful in circumstances when you have a fine-grained grid value. Even though this does not cover all the combinations, the best-performing combination will still be closer to the combination that the Grid search method can find. The major advantage of the Random search method, however, is that increasing the hyperparameter space does not affect the performance of the method.

Fortunately, both Grid search and Random search are implemented in Python using the `scikit-learn` library.

Bayesian optimization

Whereas Grid search and Random search rely on brute-force and random methods respectively, the idea of Bayesian optimization is completely different. For a given function, this method builds a probabilistic model and analyzes the model to make the best decisions about where to next evaluate the function. In this case, the probabilistic model is also referred to as the "surrogate" model. Once this surrogate model is defined, then we use that to select the next hyperparameter combination to try. An example of the surrogate model can be the Gaussian model:

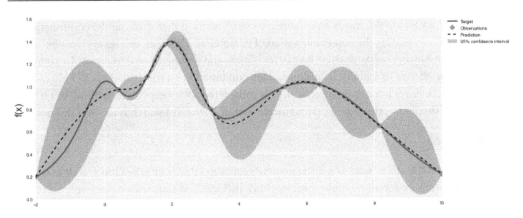

Figure 9.8 – Example of Bayesian optimization

> **Note**
> Image Credit: `https://github.com/fmfn/BayesianOptimization`

Hyperparameter tuning libraries

The main advantage of Bayesian optimization compared to the other two methods is the low computational resources requirement and that it requires only a few iterations. Let's take a look at how you can implement these hyperparameter methods.

Hyperparameter tuning is tough because hyperparameters are not always intuitive. In addition, the complexity of architecture—for example, NNs, non-convex optimization, and the curse of dimensionality—increase the computation requirements because of the large parameter space. Thankfully, we have some libraries that can enable hyperparameter tuning- KerasTuner, Ray Tune, Optuna, HyperOpt, Scikit-Optimize, Microsoft's NNI, Google's Vizer, AWS Sage Maker, and so on.

It is beyond the scope of this chapter to go into the details of each of these libraries, so let's discuss KerasTuner, which is the most popular DL library for hyperparameter tuning.

KerasTuner

KerasTuner is a hyperparameter tuning library specifically built for humans. It can be used by AI practitioners, creators, and model designers. Using Keras, hyperparameter tuning is very easy. It is mainly built for scalable hyperparameter optimization to address the pain points of hyperparameter search. One can easily configure the search space and then use the available search algorithms (Grid, Random, Bayesian, which you have learned previously) for your models. The algorithms are built, and hence it is easy for researchers to experiment to identify the best hyperparameter for their models.

Let's illustrate this with a simple example, as shown here. The following code uses the Keras library to train a simple CNN model:

```python
from keras import Sequential
from keras.layers import Conv1D, MaxPool1D, Flatten, Dense,
Dropout

model = Sequential()
model.add(Conv1D(filters=32, kernel_size=12, activation="relu",
input_shape=(50,4)))
model.add(MaxPool1D())
model.add(Flatten())
model.add(Dense(20, activation='relu'))
model.add(Dropout(0.2))
model.add(Dense(2, activation='softmax'))
model.compile(loss='binary_crossentropy', optimizer='adam',
metrics=['binary_accuracy'])
model.summary()
```

This code uses the Keras `Sequential` module to build a simple 1D CNN model for a binary classification problem, as follows:

- Input data of size 50 by 4
- 1 convolutional layer containing 32 kernels of size 12 (motif size)
- 1 max pooling layer with a pool size of 4 (filter size)
- A flatten layer to convert to flatten the nodes into a single layer
- A dense fully connected layer consisting of 20 nodes and a Relu activation function
- A dropout fraction of 0.2
- The final output layer consists of 2 nodes and a softmax activation function
- Loss is categorical binary entropy with Adam as the optimizer for the learning rate and binary accuracy as the metric for model evaluation

Now, let's see how we can use the KerasTuner library for tuning hyperparameters with the same model. Follow these steps:

1. First, wrap the model in a function.
2. Define hyperparameters that you want to optimize.
3. Replace the static values with the hyperparameters that you defined in *step 2*.

Let's see how it is all done in the following code snippet:

```
def model_builder():
    learning_rate = Choice('learning_rate', [0.01,
0.001, 0.0005], group='optimizer')
    droput_rate = Linear('dropout_rate', 0.0, 0.5, 5,
group='dense')
    numb_dimensions = Range('num_dims', 8, 32, 8,
group='dense')
    numb_layers = Range('num_layers', 1, 3,
group='dense')
    numb_filt = Range('num_filters', 8, 16, 24,
group='cnn')
    kern_size = Range('k_size', 8, 10, group='cnn')

    model = Sequential()
    model.add(Conv1D(filters=numb_filt, kernel_
size=kern_size, activation="relu", input_shape=(50,4)))
    model.add(MaxPool1D())
    model.add(Flatten())
    for _ in range(numb_layers):
        model.add(Dense(numb_dimensions,
activation='relu'))
        model.add(Dropout(droput_rate))
        model.add(Dense(2, activation='softmax'))
        model.compile(loss='binary_crossentropy',
optimizer=Adam(learning_rate), metrics=['binary_
accuracy'])
        model.summary()
        return model
```

As can be seen from the preceding code snippet, we first defined a `keras_tuning` function that was used to wrap the whole code shown previously.

4. Next, we defined the hyperparameters that we wanted to optimize, such as learning rate, dropout, number of dimensions, number of layers, and finally, number of filters. Finally, we replaced the static values in the commands with the hyperparameters that we had defined.

5. Next, we initialize a tuner (in this example, RandomSearch)

```
tuner = kt.RandomSearch(
    model_builder,
```

```
objective='val_loss',
max_trials=5)
```

6. Finally, start the search and get the best model using training data:

```
tuner.search(x_train, y_train, epochs=5, validation_
data=(x_val, y_val))
best_model = tuner.get_best_models()[0]
```

Model evaluation

This is the last step in the DL life cycle before deploying the model to production, where it will use the live data for predictions. So, in a way, model evaluation is the critical phase of the DL life cycle because the accuracy of the predictions once the model is deployed in production depends on this phase. After the model is trained, it needs to be evaluated for its performance and accuracy using key metrics such as **mean squared error (MSE)**, **root mean squared error (RMSE)**, and so on.

Generally, multiple models are trained using various methods and then evaluated for the effectiveness of each model before selecting a final deployment model, as explained previously. There are two different types of model evaluation based on the type of data—online model evaluation, which uses live data, and offline model evaluation, which uses historical data. Model evaluation in terms of performance metrics is generally related to offline models, and the model evaluation after production using live data is to ensure the models are robust, fair, calibrated, and working. For model evaluation, ideally, the input data that was used to build the model should be like the input data that the model will have to use in production, but this is not always possible because there is huge variability in the real-world data that the model sees in production. As a result, the input data in production is much noisier compared to the input data during model development. To address this, we can make small changes to the model development process. We can add some random noise to the training data during the model training process so that the algorithm can learn from the random noise, and it can then use those learnings during predictions on real-world noisy data.

The performance metrics or evaluation metrics are generally tied to the type of DL task. There are different performance metrics for classification tasks, regression tasks, and clustering tasks. Out of these, classification and regression belong to SL tasks and clustering belongs to **unsupervised learning (UL)** tasks. Let's understand each of the different types of metrics in the next section.

Classification metrics or performance statistics

Classification tasks are tasks related to predicting the class label of an observation. Binary classification tasks involve two class labels, whereas multiclass classification tasks involve more than two class labels. There are several classification metrics related to both classification types, as shown next.

Accuracy

This is the most used classification metric to measure the performance of a model. It is simply a measure of how often the classifier makes the right predictions. It is defined as the ratio of correct predictions to the total number of all the predictions, as shown here:

$$Accuracy = \frac{\#\ correct\ predictions}{\#\ total\ predictions}$$

The higher the accuracy, the better the model, and vice versa. High accuracy means the model performed well with the test data. For example, an accuracy score of 90% for a classification model indicates that 9 out of 10 times the model predicts correctly. Accuracy is easy to understand and interpret, but it is not the right metric in the case of unbalanced data where one class has more observations compared to the other class. Using accuracy for an unbalanced dataset gives you a statistic that is skewed toward the majority class because that class with more examples will dominate. For example, in general, there are more data points for a protein-coding gene compared to a pseudo gene because pseudo genes are very uncommon and so if you train a classification model on this kind of data, the model might be biased towards the majority class in this case, protein coding genes.

Even though accuracy is easy to understand, it doesn't give you the full picture of the predictions. If there are two classes—for instance, **transcription factor** (**TF**) or no TF, the accuracy gives equal preference to both classes, which sometimes is not enough. You might be interested in knowing how many of many observations don't fall into no TF versus TF because it is important to know if the cost of misclassification might differ for the two classes, or if there is an unbalanced class issue in the data. This is when you need something called a confusion matrix that shows a more detailed breakdown of classification (correct class versus not correct class). In a typical confusion matrix (*Figure 9.9*), the rows correspond to the ground truth (or original labels) and the columns correspond to observations (predictions), as illustrated in the following example:

	Positive prediction	Negative prediction
Positive label	**True Positives (TP)**	**False Negatives (FN)**
Negative label	**False Positives (FP)**	**True Negatives (TN)**

Figure 9.9 – Confusion table

Here, the following applies:

- TP: Number of positive labels correctly predicted
- TN: Number of negative labels correctly predicted
- FP: Number of negative labels incorrectly predicted as positive
- FN: Number of positive labels incorrectly predicted as negative

Now, accuracy can be written as this:

$$Accuracy = \frac{TP + TN}{TP + TN + FP + FN}$$

For example, the confusion table presented here shows a test dataset consisting of 50 examples in the positive class (TF) and 100 examples of the negative class (no TF):

	Positive prediction	Negative prediction
Positive label	40	10
Negative label	5	95

Figure 9.10 – Example of a confusion matrix

Looking at the positive labels here, the accuracy of the positive class is 40/40+10*100 = 80%, and similarly, the accuracy of the negative class is 95/95+5*100 = 95%. So, as you can see clearly, the accuracy of the negative class is much higher than the positive class, and this information is clearly lost if we calculate the accuracy for the whole data 40+95/50+100*100, which is only 90%.

Precision and Recall

These are important classification metrics that would help understand the finer details of the classification performance. Precision informs out of all the decisions made by the classifier to be relevant, how many of them are truly relevant. In contrast, recall informs out of all the relevant items in the data, how many of them are made by the classifier.

Mathematically, precision and recall are calculated as follows:

$$precision = \frac{\#\ correct\ predictions}{\#\ decisions\ made\ by\ classifer}$$

$$precision = \frac{TP}{TP + FP}$$

$$recall = \frac{\#\ correct\ predictions}{\#\ total\ relevant\ items}$$

$$recall = \frac{TP}{TP + FN}$$

Precision and recall can be used together in a metric called F1, which is a harmonic mean of precision and recall. This is an important metric to evaluate a classifier, and unlike the arithmetic mean, the harmonic mean tends toward the smaller of the two elements. So, for example, if either Precision or recall is low, the F1 score will be low, as illustrated here:

$$F1 = 2\frac{precsion * recall}{precision + recall}$$

Visualizing performance

Many classifiers allow users to change the thresholds so that they can vary how many examples get classified as positives. There are two types of curves that we can plot.

ROC

This stands for the **receiver operating characteristic (ROC)** curve, which is a measurement of the performance of a model. It is generally used to compare multiple models to identify the best-performing model. This curve shows the sensitivity of the classifier by plotting the **True Positive Rate (TPR)** as a function of the **False Positive Rate (FPR)**. The **area under the curve (AUC)** measurement indicates the performance of the classifier. Even though AUC is represented by a single numeric (between 0 to 1), the number summarizes the ROC and provides information about the classifier. A high-performing model will have an AUC measurement of 1, which never happens in the real world, and a poor-performing model will have an AUC measurement of 0.5, which represents the prediction is not better than random, as shown in the following diagram. The ROC-AUC does not depend on the distribution of classes in the dataset, and so it's a preferred metric for unbalanced data:

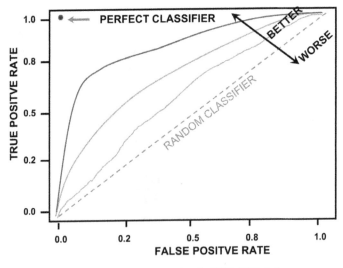

Figure 9.11 – An example of a ROC-AUC plot

The preceding AUC-ROC plot shows a few example models with varying levels of AUC. The random classifier is shown as a perfect diagonal line. The ROC above the random classifier performs better than the random classifier, and the classifiers below the random classifier show those models that performed worse than the random classifier.

Precision-recall (PR) curve

This curve plots the precision values as a function of the recall The AUC for a PR curve is a measure of the performance of the model. A perfect model will have a PR-AUC of 1.0. Unlike a ROC-AUC curve, for a fixed classifier, the PR-AUC will change depending on the ratios of classes in the dataset and so is not a recommend metric for unbalanced data.

Log-loss

This is another classification metric that is mainly employed in a classifier that produces numerical probability instead of class labels 0 or 1. Log-loss is a measurement of accuracy that incorporates the idea of probability and is equivalent to $-1 * \log(likelihood\ function)$. If we go by the preceding example, if the two classes are represented by TF and no TFs and if the classifier predicts as TF for a non-TF observation (False positive) with a probability of 0.51, even though the classifier made a mistake, it's a near miss because the probability is very close to the decision. The log-loss metric is incorporated in many Python libraries and so it's beyond the scope of this chapter to go into the mathematical details of the metric.

Regression metrics

Unlike classification tasks, a regression algorithm predicts a continuous value and so none of the preceding classification metrics would be appropriate for evaluating regression models. For example, gene expression prediction is considered a regression task because the final output is a numerical value that indicates the level of expression of an input gene. Let's understand the most popular metrics that are currently used to evaluate regression models.

RMSE

Root Mean Squared Error (**RMSE**) is the most common metric for evaluating regression models. This is defined as the root of the mean of the squared errors (predicted value versus true value):

$$RMSE = \sqrt{\frac{\Sigma(y - y')^2}{2}}$$

In the preceding equation, y represents the true value and y' represents the predicted value. One advantage of RMSE compared to MSE is that the return value from RMSE has the same units as the target variable, so it's easy to make comparisons. The higher the RMSE, the lower the performance of the model, and vice versa. One drawback with RMSE is that it does not work well on a dataset that has outliers since the metric is based on an average. Instead, we can use **median absolute percent error** (**MAPE**) which is the median of all absolute percentage errors calculated between predictions and true values.

R^2

This is the proportion of variation in the outcome explained by the predictor variables. This is mainly applicable to multiple regression models that indicate the squared correlations between the true outcome values compared to the predicted values from the model. A high R^2 score indicates a better model. To adjust the R^2 score for multiple variables in the model, people use Adjusted:

$$R2 = 1 - \frac{RSS}{TSS}$$

Here, the following applies:

RSS = sum of squares of the residuals

TSS = total sum of squares

Now, let's take all of these learnings and apply them to a real-world problem in the next section.

Use case – Predicting the binding site location of the JunD TF

In the last section of the chapter, we will see how to leverage a DNN algorithm to solve the problem of prediction of **Transcription Factor Binding Site** (**TFBS**) predictions in the human genome. We will build a DL model using one of the most popular NN architectures that are commonly used in genomics: CNNs, which we learned about previously. But before that, let's understand the problem and data in detail.

TFs are proteins that control gene regulation. They bind to the regulatory regions of the DNA such as Promoters and either promote or repress gene expression. Each TF has a specific binding motif that it binds to, which is referred to as a TFBS. The identification of a TFBS is very challenging because the binding motifs are generally very small (<10bp) or not completely specific, or a TF may bind to many similar but not identical sequences, or in some cases, some bases in the motifs are generally more important than others. Conventional methods for identifying a TFBS rely on the **position weight matrix** (**PWM**), which indicates how much preference the TF has for each base at each position of the motif. This method has limitations:

- One needs to know the relative importance of the bases in the motif
- It assumes that every position in the motif is independent of the other

- Non-uniform length of the motif
- Even though the motif is the main contributor to the binding, the regions outside of motifs are also shown to be important

In addition, other factors that control TF binding to motifs are the physical shape of the DNA, methylation, and so on. This problem calls for a sophisticated solution, and this is where DL might be a suitable approach. Let's break this use case into different phases of the DL life cycle that we have learned about previously.

Framing the TFBS prediction problem in terms of DL

Let's start by framing the business problem as a DL problem. The goal of this exercise is to build a predictive model from the data that can predict if a DNA sequence has a TFBS or not.

Processing the data

After we frame the business problem into a DL problem, the next phase is data processing, which collectively consists of data collection and data wrangling.

Data collection

For this example, we will use TF JunD (`https://www.ebi.ac.uk/interpro/entry/InterPro/IPR029823/`), which is a member of the TF **activator protein** (**AP**)-1 family and plays a central role in regulating gene transcription in various biological processes. For this hands-on exercise, note the following:

- We will use the data that has been generated by an experiment that was done to identify every place in the human genome where TF JunD binds
- To keep things manageable, only sequence data from chromosome 22 (one of the smallest) is included in the data and only a few data samples are used for this example

Data wrangling

After data collection comes the most important phase, which is data wrangling. This phase has two sub-phases: data preprocessing and feature engineering.

Data preprocessing

The data has already been cleaned, preprocessed, and transformed and is provided in NumPy format. Briefly, the full 22 chromosomes are split up into segments of 101 bases long, and each segment has been labeled to indicate whether or not it includes JunD binding sites, which are obtained experimentally. What this means is, we do not need to do any of these steps for this data. In addition, the data is available separately for train, testing, and validation purposes. Here is an example of how the whole sequence is cut into pieces of 101 bases each:

```
ATTAAAGGTTTATACCTTCCCAGGTAACAAACCAACCAACTTTCGATCTCTTGTAGATCTGTT
CTCTAAA
CGAACTTTAAAATCTGTGTGGCTGTCACTC | GGCTGCATGCTTAGTGCACTCACGCAGTATAA
TTAATAAC
TAATTACTGTCGTTGACAGGACACGAGTAACTCGTCTATCTTCTGCAGGCTGCTTACGGT | TT
CGTCCGTG
TTGCAGCCGATCATCAGCACATCTAGGTTTCGTCCGGGTGTGACCGAAAGGTAAGATGGAGAG
CCTTGTC
CCTGGTTTCAACGAGAAAA | CACACGTCCAACTCAGTTTGCCTGTTTTACAGGTTCGCGACGT
GCTCGTAC
GTGGCTTTGGAGACTCCGTGGAGGAGGTCTTATCAGAGGCACGTCAACA | TCTTAAAGATGGC
ACTTGTGG
CTTAGTAGAAGTTGAAAAAGGCGTTTTGCCTCAACTTGAACAGCCCTATGTGTTCATCAAACG
TTCGGAT
GCTCGAAC | TGCACCTCATGGTCATGTTATGGTTGAGCTGGTAGCAGAACTCGAAGGCATTCA
GTACGGTC
GTAGTGGTGAGACACTTGGTGTCCTTGTCCCTCATGTG | GGCGAAATACCAGTGGCTTACCGC
AAGGTTCT
TCTTCGTAAGAACGGTAATAAAGGAGCTGGTGGCCATAGTTACGGCGCCGATCTAAAGTCATT
TGACT | TA
```

Feature engineering

For this dataset, we will rely on DL to extract features automatically and use them in the model training. The only feature engineering step is to represent sample features with one-hot encoding. Let's understand how this can be done.

Since the DNA sequence is a non-numeric data type, to extract features from this DNA sequence, we must do the following:

- For DNA, we have 4 bases or categories A, C, T, and G, and for each base, we will have 4 numbers; the base is set to 1 and the rest is set to 0. Here is an example:

$$A = [1, 0, 0, 0]$$
$$C = [0, 1, 0, 0]$$
$$T = [0, 0, 1, 0]$$
$$G = [0, 0, 0, 1]$$

- So, the sequence of ATGCGT can be represented as follows:

$$[[1, 0, 0, 0],$$
$$[0, 0, 1, 0],$$
$$[0, 0, 0, 1],$$
$$[0, 1, 0, 0],$$
$$[0, 0, 0, 1],$$
$$[0, 0, 1, 0]]$$

- Ultimately, for each sample, we have a feature vector of size (number of bases, one-hot encoding of each base). In this example, the feature vector will be (101, 4).

Now that we finished looking at the samples, let's look at the target variable. Sample labels are indicated by a single number. 0 doesn't contain the binding site, and 1 contains the binding site. The data, as mentioned before, has already been processed by the experimental authors and stored in NumPy (.npy) format. We can use libraries such as NumPy, pandas, and others to quickly look at the data:

1. After you download the data, load the requisite libraries and then set the training data path corresponding to the training data:

```
# Load required libraries
import os
import numpy as np
import pandas as pd
# Set the path to the direction where the data is
downloaded
path = "JunD_TF"
os.chdir(path)
```

```
## Set the training data path
train_path =  path + "/train"
## Load the train data numpy array into a dictionary for
the
train = {}
with open(train_path+'/train_X.npy','rb') as f:
    train['X'] = np.load(f)
with open(train_path+'/train_y.npy', 'rb') as f:
    train['y'] = np.load(f)
```

2. Let's look at the size of the training data:

```
train['X'].shape
(276216, 101, 4)
```

This shows that the training feature matrix consists of 276,,216 samples, each containing a DNA sequence of size 101bp and a one-hot encoding matrix of length 4:

```
train['y'].shape
(276216, 1)
```

3. This shows that the target variables for the training data consist of 276,216 samples. Let's look at the target variables and the number of samples corresponding to those target variables:

```
pd.DataFrame(train['y']).value_counts()
0      275048
1        1168
```

The target variables, as discussed before, contain two classes—class labels 0 and 1. Class label 0 here indicates that the same does not contain TFBS and class label 1 indicates that it does contain TFBS.

As we can see, class 1 has a smaller number of data points compared to class 0. Only 1,200 sequences contain the TFBS to our DNA sequences (samples) in the training set. But we will try adjusting for this target imbalance while building the model.

Model training

Now comes the most interesting part of the DL life cycle, which is training the model. For this, we will use the training data for model training and the validation data for validating the model and reporting out the metrics while we train it. In addition to these, for the final model evaluation, we will use the test data, which is kept separate and not used during model training. So, what do we need for the model training?

1. Firstly, we will build a simple binary classifier, as we are interested in only a two-way outcome: whether a DNA sequence contains a binding site or not.

2. Secondly, even though model selection—as indicated previously—is one of the key aspects of the DL life cycle, and in theory (and in practice too) we could take the same problem and try to solve it in multiple ways using multiple NN architectures such as MLP, CNN, RNN, and so on or as discussed earlier, understand the problem domain and select the DNN architecture that suits the problem, for this use case, we will be using CNN since it is very easy to implement and use it. In addition, there is a lot we can do with CNNs, as learned in the *Chapter 5, Introducing Convolutional Neural Networks for Genomics* but for the current tutorial, we will keep our model simple and don't complicate things. We will use Conv1D, a subclass of the popular Conv2D class (mainly used with image data). Thirdly, as with any NN, we'll need to set an evaluation metric we'll monitor during training, a loss function, and an optimizer. Since this is a classification problem, we will use classification metrics as seen before, such as AUC score and ROC-AUC curve. For the loss function, we will use binary cross-entropy, which is commonly used for binary classification problems.

3. Finally, for the optimizer, we'll use SGD, which is commonly used as an optimizer as well.

4. Let's see all of this code first using the Keras library:

```
from tensorflow.keras.layers import Conv1D, Dense,
MaxPooling1D, Flatten, Dropout
from tensorflow.keras.models import Sequential
model = Sequential()
model.add(Conv1D(filters=15, kernel_size = 10,
padding="same",
                 input_shape=(train['X'].shape[1], 4)))
model.add(MaxPooling1D(pool_size=2)),
model.add(Flatten())
model.add(Dropout(0.02))
model.add(Dense(16, activation='relu'))
model.add(Dense(1,activation='sigmoid'))
model.compile(loss='binary_crossentropy',
optimizer='adam',
```

```
                              metrics=['binary_accuracy'])
    model.summary()
```

5. Let's train the model next:

```
callback = [EarlyStopping(monitor='val_loss',
patience=2),
              ModelCheckpoint(filepath='best_model.h5',
monitor='val_loss', save_best_only=True)]
history = model.fit(x=train['X'],          # feature
matrix
                      y=train['y'],              # target
vector
                      validation_data = (valid['X'],
valid['y']), # validation data
                      callbacks=[callback]   # Early stopping
                      )
```

Let's look at some metrics from the trained model:

```
history.history
{'loss': [0.022289589047431946], 'binary_accuracy':
[0.99615877866745], 'val_loss': [0.03112887032330036],
'val_binary_accuracy': [0.9956555962562561]}
```

All these metrics are super good and indicative of a well-trained model.

6. Let's evaluate the model on the test data that we have and see how well it performed on unseen data:

```
results = model.evaluate(test['X'], test['y'], batch_
size=128)
print("test loss, test acc:", results)
test loss, test acc: [0.03267550468444824,
0.9956846833229065]
```

Again, the metrics on the test data are very promising, showing very low loss and very high accuracy on the test data.

The AUC of both training and validation data is very close, which is what we wanted, and we have an improvement in model performance, reaching a validation AUC of 0.76.

Summary

Understanding the key concepts of the DL life cycle—which involves several phases, right from understanding the business problem, all the way to model monitoring—is important for developing DL models for genomic applications because each of these phases is critical for the success of building a highly accurate model. Developing DL models for genomic applications not only involves conceptual understanding but also knowing how to practically apply these algorithms in genomics using the available DL libraries.

We started the chapter by going through the iterative steps of a DL life cycle which start with understanding the business problem and culminating with model monitoring. We understood how DL can help to solve a business problem with the right framing of the business problem into a DL problem. Fortunately, the DL community has come up with detailed steps that can help make this process easy.

Since training, tuning, and evaluation are the main topics of this chapter, we spend quite a bit of time looking at those three key steps in the DL life cycle. Training data and model algorithms mainly decide the fate of the model-building process. For example, if the training data is bad, you will end up building an inaccurate model because the model is as good as the data that gets fed into it during the model training process. So, it's worth curating the data and making training data that will make the models learn something useful. We understood how model development is not a linear process but an iterative process.

Models are not going to be successfully built in the first iteration of training, and it requires optimization of hyperparameters through a process called hyperparameter tuning. This process is very complicated because the process is not intuitive. We have seen several examples of automated hyperparameter libraries to make this process easier.

Finally, the last step of this chapter is the model evaluation which ultimately decides the fate of the DL model: whether it is going to be deployed in the production environment or needs more training. In that section, we have seen several metrics that are currently available to evaluate the DL models and ensure the model is generalized very well on unseen data. Lastly, we took all of these basics of the DL life cycle and applied them to a real-world use case which is TFBS predictions using CNN architecture.

Now that you have looked at the first few phases of the DL life cycle, in the next chapter, you will learn about interpreting the built model.

10
Model Interpretability in Genomics

Deep learning (DL) methods have been widely adopted in genomics for extracting biological insights and model predictions because of their superior performance in predictions and classification tasks through their **deep neural network** (DNN) architecture. Even though the accuracy and efficiency of these model predictions are the primary goals of DL in genomics applications, the decisions made by these DNNs is also important in genomics toward the goal of understanding cellular and molecular mechanisms. In **Machine Learning** (ML) and DL, "Model interpretability" refers to how easy it is for humans to understand the decisions made by the model. The more interpretable the models are, the easier is it to understand the model's decisions. In contrast, difficulties in model interpretation limit the practical utility of DL models and reduce confidence in their adoption. However, it's not easy to interpret DL model behavior in a way that humans can understand because of the features they use to draw conclusions and the mathematical calculations they use and hence they are referred to as black-box models. It is important to understand *why* a model has made decisions before it is put into production, especially in domains such as life sciences, biotechnology, genomics, medicine, healthcare and so on.

Model interpretability is one of the hottest areas of DL research and is a complicated subject. By interpreting the model's behavior, we can fill in the gaps in our understanding of how models work and increase the trust in the models that are built using DL. Model interpretability can be used to discover the knowledge of the model, justify the model predictions, and improve the model. In addition, model interpretability is often a key factor when a model is used in the product, a decision process, or research. However, model interpretability is not easy to perform, especially with DL models, where it is even harder because of the complexity of the model, leaving few clues for researchers about the underlying behavior of these models. Thankfully, several model interpretability methods and tools exist that can help researchers to help interpret model behavior. In this chapter, you will be introduced to gentle concepts in DL model interpretability. Model interpretability introduced here helps genomic researchers and other scientists to understand the business context of building models and using the same for predictions. By the end of the chapter, you will have learned what model interpretability is and an introduction to the tools for performing model interpretability to uncover biological insights from genomic datasets. As such, here is the list of contents that will be covered throughout the rest of the chapter:

- What is model interpretability?
- Unlocking business value from model interpretability
- Model interpretability methods in genomics
- Use case – Model interpretability for genomics

What is model interpretability?

Deep learning's popularity is mainly because of the sophisticated algorithms such as DNNs it uses to perform complex tasks. If trained popularly, the models are not only accurate but also generalize very well on real-world data. DL, with its ability to extract novel insights using automatic feature extraction and to identify complex relationships in massive datasets showed superior performance compared to the state-of-the-art conventional methods. The promise of these DL models, however, comes with some limitations. With their black-box kind of nature, these DL models face problems in explaining the relationship between inputs and predicted outputs or, in other words, *model interpretability*. It is humanly impossible to follow the reasoning for a particular prediction using these black-box models. You might be wondering if the DL models perform and generalize well, why you would not trust the model that it is making the right decisions. There can be two possible reasons:

1. Firstly, as we saw in *Chapter 9, Building and Tuning Deep Learning Models*, model evaluations are mainly done using few model metrics, which is an insufficient to evaluate how models work on real-world tasks.

2. Secondly, because these DL models are currently being used for high-stakes decision-making in many domains including healthcare, medicine, pharmaceuticals, and so on, there must be much clarity on how these predictions or why a particular decision is made by the model to build trust.

There is a lack of understanding of these concepts compared to how data is entered and how final choices are made. No matter how accurate these models are and how advanced our understanding of biological mechanisms is, how can we trust these models if we don't understand how they work?

Black-box model interpretability

Before we discuss model interpretability as it relates to DNNs, let's first understand the concept. Most **machine learning** (**ML**) models are interpretable, which means it's easy for humans to understand why the model made those predictions. For example, the simplest ML model is linear regression, in which the final output prediction Y is the weighted sum of all the features from the input X and the bias term. The easiest way of understanding the linear regression model is to look at the coefficients (weights) that are learned during model training for each feature. These coefficients indicate how much the model output changes when 1 unit of values of input features, whereas, in the case of DNNs, they are generally referred to as "black-box" models because of their complex architectures and hard-to-interpret model decisions. So why is it hard to interpret a DNN model prediction?

Here is an example of a simple fully connected **neural network** (**NN**) architecture that you have seen multiple times before (*Figure 10.1*). Even with this simplest of NN architectures, it's hard to understand which feature is contributing to the final model output because of several layers and multiple nodes per layer:

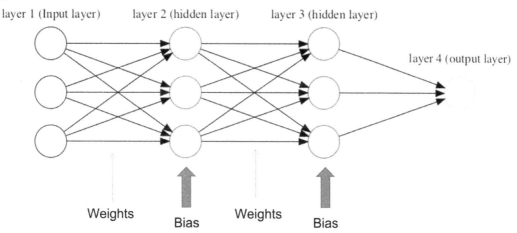

Figure 10.1 – A simple two-layered NN

If we cannot explain how the model predictions are made with this simplest of NN architectures, now imagine we have a DNN with millions of neurons and several hundred layers, as shown here:

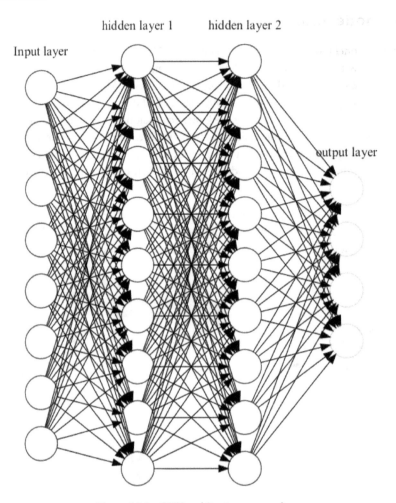

Figure 10.2 – DNN architecture example

Interpreting a model with this complicated NN architecture as shown above is even more challenging and humanly impossible.

The challenge in model interpretability faced by current DL models applies to the field of genomics as well because several genomic applications rely on DNNs for predictions and classification-related tasks, as seen before. For example, in the previous chapter, we built a DL classification model for **transcription factor binding sites** (**TFBS**) predictions. We can now use the model on an unknown DNA sequence to predict if it contains TFBS or not, but we don't know *how* and *why* these predictions are made by the model. This is where model interpretability is helpful. Model interpretability algorithms will not only help with interpreting the model decisions but also augment at several points in the DL life cycle phases such as data collection, model training, and model monitoring, as seen in the following diagram:

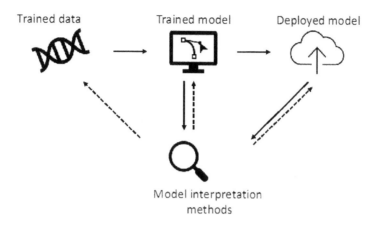

Figure 10.3 – How can interpretation methods help in a typical DL life cycle?

At each of the three stages shown in *Figure 10.3* (data collection, model training, model deployment), current model interpretability methods can help address the problem of model interpretability.

Unlocking business value from model interpretability

When it comes to predictive modeling, there is a trade-off between wanting to know *what* is predicted or *why* a prediction was made, or you do not care why a decision was made. Each of those scenarios depends on the use case. For example, if you are building a model for **research and development (R&D)** purposes, then you would sacrifice model interpretation over accuracy, whereas if you are building a model for business use, it is important to understand how the model is making that decision. Either way, it is important to know the model's behavior not only to understand why some decisions were made but also for debugging and model improvement.

Let's now understand why model interpretability for DL is important and how it helps to unlock practical business benefits such as better business decisions, building trust, and increasing profitability in the following section.

Better business decisions

Before model deployment, the models are evaluated on a test dataset, as seen in *Chapter 9, Building and Tuning Deep Learning Models*, using one or more evaluation metrics such as accuracy, AUC, and so on. When a company decides to deploy this model as an application to use real-world data, relying on this single metric may lead to new questions such as "Which of the input features are impacting the final model performance and predictions?", "Can those predictions be trusted with real-world data?", or "Can I trust my model since it is performing exceptionally well?" and so on. If this model is deployed for non-risky tasks, then the penalty for making a wrong prediction is low, but if the model is deployed for business use cases, focusing on just the evaluation metrics is not enough. We must understand model decisions for reasons such as model improvement, debugging, trust, compliance, and so on.

Let's understand this through an example.

It is well known that current cutting-edge targeted therapy to treat Cancer, even though works great, suffer from several limitations such as laborious, expensive, time consuming, and so on. DL is a great method to address this limitation and surpass current state-of-the methods in drug response predictions because they capture intricate biological interactions to the specific drug.

For example, if a company is interested in producing a drug to treat cancer and based on their research, they find that "undruggable proteins" such as **transcription factors (TFs)**, which are involved in transcription mechanisms, can be potential targets for cancer drug development or target therapy. They now collect lots of genomic data from several genomic databases such as the **Gene Expression Omnibus (GEO)**, **The Cancer Genome Atlas (TGCA)**, the **Cancer Cell Line Encyclopedia (CCLE)**, pharmacogenomics database and so on. Then, they trained a DNNs.

The following diagram provides a visual representation of such a DNN model:

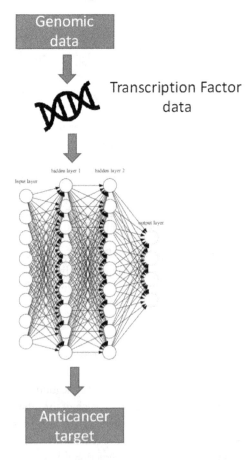

Figure 10.4 – A simplified illustration of a DNN model for identifying anticancer targets from genomic data

Before they use the trained model to predict patient drug response and then develop anticancer drugs based on those predictions, they have three possible routes:

- They might be interested in knowing the probability of how effective the developed anticancer drug is for the patient

- They care about model interpretability as to why the model made that decision/prediction

- They do not care why a decision was made or if it made the right decision

Among the three scenarios, knowing *why* made that prediction can help them learn more about the problem, the data, and the reason. If the model is mainly used for R&D and a mistake may not have serious consequences, then you probably don't care about the model's decisions. In general, model interpretability can lead to an improvement in the model. Model interpretability can sometimes lead to the identification of outliers which can either lead to potential revenue opportunities or potential liabilities waiting to happen and both help companies help prepare and strategize decisions.

Building trust

For organizations, trust is their reputation and is something that cannot be bought with money. The goal of model interpretability is to make models better at making decisions. If the models fail, at least they should explain why they would have failed. So, having good model interpretability is key for a model to build trust among the company and the stakeholders involved. There is no substitute for that.

Profitability

If model interpretability ensures better decisions, builds trust, and improves model predictions, then this will lead to its increased use and enhance overall brand reputation, and ultimately increased financial revenue. In contrast, if there are concerns with any of those issues, they will adversely impact both profits and reputation.

Model interpretability methods in genomics

The field of genomics has garnered so much attention lately because of the advances in high-throughput methods, such as **next-generation sequencing** (**NGS**), and other omics technologies such as proteomics, metabolomics, and so on. This has resulted in abundant data such that researchers are in a dilemma about how to use this. The DNN methods showed superior performance compared to the state-of-the-art conventional methods in many genomics applications in medical research, especially in imaging tasks, tumor identification, antibody discovery, motif finding, genetic variant detection, and chromatin interaction, to name a few. However, the major complaint from DNN architectures is that they are black-box models. What that means is that we don't know how these models made decision on a given dataset. To make predictions with a DL model, the input data is passed through several layers of a DNN, each layer containing several nodes that have learned weights and activation functions.

This is a complicated process that involves millions of mathematical operations, and humans can't comprehend the exact mapping from input data to prediction. To provide insights into DNN model behavior, researchers have come up with interpretability methods.

Several methods and tools for model interpretability exist for DNNs so that humans can easily understand them. These methods visualize features and concepts learned by the DNN during training, explain individual predictions made by the model, and simplify DNN model behavior. These model interpretability methods can be broadly divided into *global* and *local*. Global methods aim to understand how a model decides overall, whereas local methods do it for a single instance. There are several methods currently available for both types, with the goal of model interpretability and improving models. Let's look at some of the model interpretability methods for DNN now.

Partial dependence plot

In the linear regression model, we have seen how model coefficients or weights can be used to understand model behavior. However, coefficients are not a good way of measuring feature importance to interpret models, mainly because the coefficients rely on the value of the input features. The more realistic measure of feature importance for a model is to understand how changing the feature impacts the output of a model. A **partial dependence plot** (**PDP**) is a way to address that (*Figure 10.5*). A PDP shows how each variable or predictor affects the model predictions. This helps researchers to make sense of what happens when various features are adjusted.

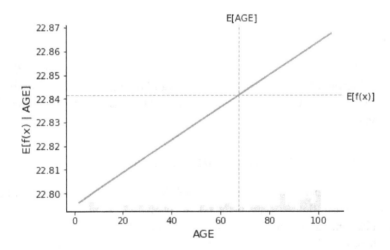

Figure 10.5 – A simple PDP illustrating the feature importance of a simple model

In the preceding diagram, the gray horizontal line shows the expected value of the model when applied to the dataset, while the gray vertical line shows the average value of a particular feature (in this example, the age of the patient). The blue line represents the average model output when we fix a particular feature (in this example, the age of the patient) to a given value. The blue line always passes through the intersection of the two gray lines.

Individual conditional expectation

The **individual conditional expectation** (ICE) method differs from PDP in that, instead of plotting an aggregate effect of all samples, ICE displays the dependence of the prediction on a feature for each sample separately in the plot (*Figure 10.6*). This is more useful because it shows the minor variations of all samples when one chose to test on a particular feature of interest.

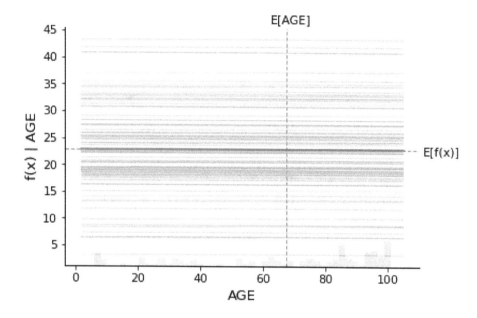

Figure 10.6 – A simple ICE plot illustrating the feature importance of a simple model

Going by the preceding example, the ICE plot shows the feature importance of a particular feature, but instead of showing the average of all the samples, it shows the dependence of the prediction for each sample separately. Even though ICE gives the more granular level of details about the model predictions and how they related to each sample, it is advisable to try both PDP and ICE for model interpretability

Permuted feature importance

Permuted feature importance (**PFI**) is a neat method for model interpretability. The idea is simple. For each feature, PFI measures the impact of the shuffling of the feature on the final model prediction error. If the shuffling of the feature increases the model prediction error, then that feature is deemed important, and if not, the feature might be not that useful for the model performance. This is because if that feature is important for the model prediction, a slight change in that feature would affect the model. In contrast, the low-impact feature doesn't affect the model prediction despite shuffling. Let's understand this using an example:

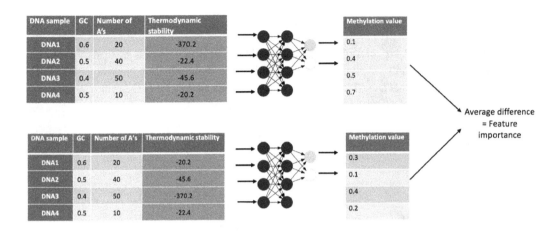

Figure 10.7 – An example of how to calculate PFI for one feature

In the preceding diagram, the table on the top left shows the three features from the original table and the predictions made. Then, we shuffle the **Thermodynamic stability** column, and then the corresponding prediction is calculated. Then, we take the average of the two model predictions to calculate the PFI for the permuted feature (in this case, thermodynamic stability). If the model relies on thermodynamic stability, then the feature importance will be very high, and in contrast, if the model doesn't rely upon it, then shuffling does not affect the feature importance. We keep repeating the same problem for the rest of the features.

Global surrogate

This is an interpretable model that is trained to approximate the prediction of the full black-box model. First, you take the trained model and make predictions on a test dataset and then train an interpretable model such as a simple ML model such as Random Forest son a prediction made by the black-box model (*Figure 10.8*). This newly trained interpretable model becomes a surrogate model for the DNN model and now we just need to interpret the surrogate model:

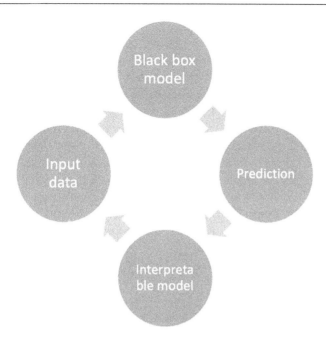

Figure 10.8 – Global surrogate method illustration

However, this method introduces additional error because it uses an interpretable model as a proxy for the black-box model and this method can only predict black-box models but cannot predict the data directly.

LIME

Local Interpretable Model-agnostic Explanations (**LIME**), as the name suggests, is a local model interpretability method that explains the model behavior for a single instance. Unlike the global method, it cannot explain the whole model. It trains interpretable models such that it approximates individual predictions.

Shapley value

Shapley is one of the most popular model interpretability methods for understanding how the features of the model are related to the outputs. This method comes from coalitional game theory, where each feature is a "player" and the prediction is the "payout". The idea of Shapley is to use game theory to interpret the model. Shapley values indicate how fairly the payout is distributed among the features (players) and hence are good for interpreting models.

ExSum

ExSum (short for **Explanation Summary**) is a recent mathematical framework to formally quantify and evaluate the understandability of explanations for models. The main advantage of ExSum compared to other methods is that this method can help provide insights about model behavior not just on a handful of individual explanations that other methods might miss. This method can provide a detailed picture of the entire model's behavior that we didn't know previously. ExSum tests a rule across the whole dataset instead of a single instance using three metrics—coverage, validity, and sharpness. The coverage metric indicates how broadly the rule may be applicable over the entire dataset, the validity metric signifies the percentage of specific examples that agree with the rule, and sharpness indicates how accurate the rule is. Simply put, we can use ExSum to see if a particular rule holds up using the aforementioned three metrics. ExSum is very powerful, and if a researcher seeks a deeper understanding of how the model is behaving, they can use it to test specific assumptions.

Saliency map

A saliency map is a frequently used method for measuring the nucleotide importance in a sequence. As a genomic researcher or a data scientist working on genomics data and interested in motifs on a DNA sequence, it is natural to think about which parts of the sequence are most important for the classification of the sequence. For instance, for the model we trained in the *Chapter 9, Building and Tuning Deep Learning Models* to predict a TFBS, it is important to understand why the model made some decisions that way. Saliency maps are one way of visualizing those patterns. Saliency maps show how the response of an output variable changes with a small change in the inputs (in this case, nucleotide sequence). These saliency maps are extremely useful visualization tools that can help us to understand the binding sites of a particular protein (in this example, TF), as shown in *Figure 10.9*:

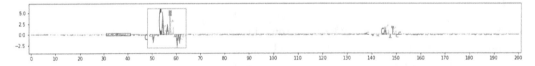

Figure 10.9 – An example of how to visualize scores on individual sequences

The preceding plot shows the ground-truth location of the motifs generated using saliency maps.

Use case – Model interpretability for genomics

In this hands-on exercise section, we will build a similar **convolutional NN (CNN)** model that we built in *Chapter 9, Building and Tuning Deep Learning Models*, but unlike in *Chapter 9*, here we will use a simulated dataset of DNA sequences of length 50 bases (whereas in *Chapter 9*, we have DNA sequence of length 101 bases). In addition, the binding sites in this example are not just for **Transcription Factors (TFs)** but any protein. The labels are designated as 0 and 1, corresponding to positive and negative binding sites (0 = no binding site and 1 = binding site).

The goal of this is to train a CNN model to predict the DNA binding site of the protein and visualize it in the predictions. Since these are artificial sequences, we have injected the AAAGAGGAAGTT motif into the positive sequence, but don't worry—the CNN doesn't know that.

Data collection

For this hands-on tutorial, we will use the simulated data (code not shown) in which we injected motifs treated as positive sequences and the rest as negative sequences. Follow the next steps:

1. Load all the necessary libraries, shown as follows:

```
from keras.layers.convolutional import Conv1D,
MaxPooling1D
from keras.layers.core import Dense, Flatten
from keras.models import Sequential
from keras.callbacks import EarlyStopping,
ModelCheckpoint
import numpy as np
Import pandas as pd
from sklearn.metrics import roc_curve, auc, average_
precision_score
from sklearn.model_selection import train_test_split
from sklearn.preprocessing import LabelEncoder,
OneHotEncoder
```

2. Load the sequence data, shown as follows:

```
input_fasta_data = pd.read_table("sequences_mod.txt",
header=None)
input_fasta_data.rename(columns={0: "sequence"},
inplace=True)
sequence_length = len(input_fasta_data.sequence[0])
```

Let's move on to the next step.

Feature extraction

The next step will be to convert the sequences to a format that the CNN or any other DL model can accept, which is one-hot encoding. Since we have learned about that already in *Chapter 9, Building and Tuning Deep Learning Models* and a few other chapters, we will skip the explanation here.

Let's see how we can leverage the available `sklearn` package for doing the same:

```
iec = LabelEncoder()
ohe = OneHotEncoder(categories='auto')
seq_matrix = []
for sequence in input_fasta_data.sequence:
iecd = iec.fit_transform(list(sequence))
iecd = np.array(iecd).reshape(-1, 1)
ohed = ohe.fit_transform(iecd)
seq_matrix.append(ohed.toarray())
seq_matrix = np.stack(seq_matrix)
Print(seq_matrix.shape)
(2000, 50, 4)
```

Target labels

So, now we have the input features ready, let's go ahead and collect the labels or targets from the same source:

```
labels = pd.read_csv('labels.txt')
Y = np.array(labels).reshape(-1)
Print(Y.shape)
(2000,)
```

Train-test split

Before we train the model, let's split the dataset into training and test datasets so that we can use training data for model training and test data for model evaluation:

```
X_train, X_test, Y_train, Y_test = train_test_split(seq_matrix,
Y, test_size=0.25, random_state=42, shuffle=True)
X_train.shape, X_test.shape, Y_train.shape, Y_test.shape
print(X_train.shape)
print(Y_train.shape)
(1500, 50, 4)
(1500,)
```

We'll now reshape the data so that it corresponds to the input format of Keras:

```
X_train_reshaped = X_train.reshape((X_train.shape[0], X_train.
shape[1], X_train.shape[2], 1))
X_test_reshaped = X_test.reshape((X_test.shape[0], X_test.
shape[1], X_test.shape[2], 1))
print(X_train_reshaped.shape)
print(X_test_reshaped.shape)
(1500, 50, 4, 1)
(500, 50, 4, 1)
```

Creating a CNN architecture

Now that we have the training data ready, let's build a CNN model and train it first. We will use the DL library Keras.

We will start training with a fairly simple CNN model with a single convolutional layer followed by a single max pooling layer, and then finally, a single dense layer. We also have an output layer consisting of 1 node corresponding to the model predictions (0 or 1):

```
model = Sequential()
model.add(Conv1D(filters=32, kernel_size=12,
input_shape=(50, 4)))
model.add(MaxPooling1D(pool_size=4))
model.add(Flatten())
model.add(Dense(16, activation='relu'))
model.add(Dense(1, activation='sigmoid'))
model.compile(loss='mean_squared_error', optimizer='adam')
```

Model training

Now that we have finished compiling the model, we are ready train the model. We feed the model with our training data (features and labels), define a batch size since we are training in the batch model, and set the number of epochs. 10 epochs would give us enough results to evaluate, and we can then ramp up if there is room to improve.

Let's quickly run it and see how well we did the model training:

```
callback = [EarlyStopping(monitor='val_loss', patience=2),
ModelCheckpoint(filepath='model_int_genomics.h5', monitor='val_
loss', save_best_only=True)]
history = model.fit(X_train_reshaped, Y_train,
```

```
batch_size=10, epochs=10,
validation_data=(X_test_reshaped, Y_test))
```

You will notice two other things here: `callback` and `history`. `callback` is mainly to ensure the model stops once it finds a good stopping point, and then the `history` variable mainly serves to monitor the loss and accuracy of the training and validation datasets during the training process:

Next, we will plot the training and validation loss to visualize how well we did with respect to model training

```
plt.plot(history.history['loss'])
plt.plot(history.history['val_loss'])
plt.title('model loss')
plt.ylabel('loss')
plt.xlabel('epoch')
plt.legend(['train', 'validation'])
```

Let's visualize the model loss plot:

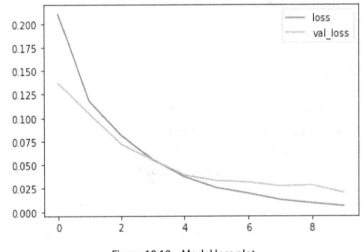

Figure 10.10 – Model loss plot

The model loss plot shows that even though there are 10 epochs, the model has done really well by keeping the training loss low at lower epochs, and the validation plot also looks great.

Model evaluation

Now that the model is trained, we can evaluate the model against the test data. Do note that the predictions are between 0 and 1 because of the sigmoid activation.

Here's what we do for calculating model metrics for model evaluation:

- We get the function of the last layer in the NN:

```
pred = model.predict(X_test_reshaped, batch_size=32).
flatten()
print("Predictions", pred[0:5])
```

- We can also calculate the AUC to get an estimate of how well the NN actually learned:

```
fpr, tpr, thresholds = roc_curve(Y_test, pred)
print("AUC", auc(fpr, tpr))
print("AUPRC", average_precision_score(Y_test, pred))
16/16 [==============================] - 0s 952us/step
Predictions [0.9709498 0.00785227 0.97798 0.99865
0.00150004]
AUC 0.9986542559156667
AUPRC 0.9985575535116882
```

Model interpretation

This is what you all have been probably waiting to see how well the model made those decisions. For this, we will generate saliency maps or motif plots, which you have learned about before.

Let's make a motif plot, as follows:

1. Let's pick a sample from the test data:

```
plot_index = np.random.randint(0, len(X_test), 1)
```

2. Get the Y ~ X for the chosen sequence that we want to plot:

```
seq_matrix_for_plotting = seq_matrix.reshape((seq_matrix.
shape[0], seq_matrix.shape[1], seq_matrix.shape[2], 1))
[plot_index, :]
plotting_pred = model.predict(seq_matrix_for_plotting,
batch_size=32).flatten()
```

3. Calculate the gradient—generate a new set of X where for each sequence, every nucleotide is consecutively set to 0:

```
tmp = np.repeat(seq_matrix_for_plotting, 50, axis=0)
a = np.ones((50, 50), int)
np.fill_diagonal(a, 0)
b = np.repeat(a.reshape((1,50,50)), seq_matrix_for_
```

```
plotting.shape[0], axis=0)
c = np.concatenate(b, axis=0)
d = np.multiply(tmp, np.repeat(c.reshape((tmp.shape[0],
50, 1, 1)), 4, axis=2))
# Calculate the prediction for each sequence with one
deleted nucleotide
d_pred = model.predict(d, batch_size=32).flatten()
# Score: Difference between prediction and d_pred
scores = np.reshape((np.repeat(plotting_pred, 50) - d_
pred), (len(plot_index),50))
```

4.   Make the actual plot:

```
import motif_plotter
import matplotlib.pyplot as plt
for idx in range(0,len(plot_index)):
fig=plt.figure(figsize=(18, 5), dpi= 80)
ax=fig.add_subplot(111)
motif_plotter.make_single_sequence_spectrum(ax,
seq_matrix_for_plotting[idx].reshape((50, 4)),
np.arcsinh(scores[idx]).reshape(50,1))
plt.show()
```

Running the preceding command will generate a saliency map, as shown here:

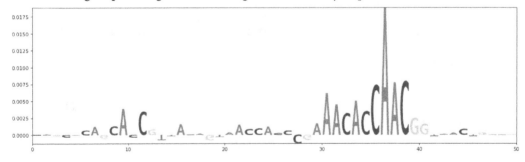

Figure 10.11 – Saliency map for bases of the positive sequences

The preceding saliency plot should show values for the AAACACCAACGG bases in the DNA sequence of one of the positive samples. Unfortunately, this is not exactly the motif that we inserted in the positive sequence. Our ground truth is AAAGAGGAAGTT This may have happened because we haven't optimized the model enough or chosen the right model for training this data. This is where we can spend a lot of time doing optimization and model interpretation. Testing different hyperparameters is a good practice that may help to get a clear signal from this data.

Summary

Model interpretability is a relatively new area, with most publications happening in the last few years, but it is a very active area of research in DL now that is of utmost importance to realize the promise of precision medicine. The ability to interpret model decisions or predictions has several business advantages and can ultimately lead to higher profits. Because of model interpretability, more and more companies are leaning toward using DL models in their decision-making processes. This is not restricted to low-risk sectors but also high-risk sectors such as medicine and genomics too. If they are not currently using model interpretability, they plan to incorporate it into their future strategy.

This chapter is an attempt to introduce you to model interpretability, why it is important, why business organizations care about it, and different methods for performing model interpretability, specifically for black-box models such as DNNs in the genomics field. The chapter started with the basics of what is model interpretability, why we care, and what value model interpretability brings to the business. With the advances in the model interpretability method, currently, there are both model-agnostic and black-box-specific models proposed to explain black-box models and we have looked at several of them broadly classified into global and local methods. Each of the methods has its advantages and disadvantages. General guidance is provided to pick the right method for model interpretability. The key takeaway from those methods is that the features used by the black-box model can be interpreted, which is important for biological domains. Finally, we looked at how to interpret models using protein binding site predictions as an example When business organizations incorporate model interpretability into their DL life cycle, it's an iterative cycle, and it results in higher profitability. For non-profits, NGOs, and research organizations, profits might not be a motive. Still, it is useful for them to keep the preceding things in mind to avoid inaccurate decision-making—reputation is at stake. Now that we have understood model deployment and hopefully improved the model, we are ready to deploy the model in production in the next chapter.

11

Model Deployment and Monitoring

The primary goal of **Machine Learning (ML)** and **Deep Learning (DL)** is to build predictive models and get insights from the data to help solve business problems. However, a trained model can only do that once you take your model and turn it into a production environment – a process referred to as **model deployment**. A deployed model enables other researchers (either within your organization or outside) to interact with and extract the most value out of it. Many models don't end up in the production environment purely because of technical challenges. Once a model is put into the production environment, constant attention is required on an ongoing basis to detect any drifts or anomalies for the success of a DL project – a process referred to as **model monitoring**. Model deployment and model monitoring are two of the most important challenges that a lot of researchers face after they build their models.

The skills and expertise that are required to build great models may not necessarily be the same skills needed to put these models in the production environment and monitor them. A common complaint you hear from data science teams is that putting models into production is a complicated process. That is the main reason that many companies and organizations have **machine learning engineers (MLEs)** to perform these types of tasks. But it is important to have the knowledge and ability to deploy and monitor the models for those who are building **Deep Learning (DL)** models. In this chapter, you will learn how to take the model you built on notebooks (for example, Google Colab) and put it into production using public open source tools such as Streamlit and Hugging Face. This chapter also describes how model monitoring is essential for the success of businesses and how to monitor models using advanced tools. By the end of this chapter, you will understand how to develop a web app using Streamlit, how to deploy the model using Hugging Face, and how to monitor models using advanced tools.

As such, the contents of this chapter are as follows:

- Introducing model deployment
- Monitoring models using advanced tools

Technical requirements

In this chapter, we will discuss some tools and software that are essential for model deployment and model monitoring. Let's go over the technical specifications.

Streamlit

Streamlit (`https://streamlit.io/`) is an open source framework for building web apps based on Python. It is a faster way to build and share web apps in minutes using Python, which fits in very well with DL frameworks such as Keras. The main advantage of Streamlit compared to other frameworks is it is easy to build, quick to deploy a model in a cloud environment, and previous knowledge of the frontend is not required. It is best suited for data scientists or genomic researchers who are not web developers and they don't need to spend their time learning web development to build apps. Streamlit enables them to quickly build a web app and share it with their collaborators, which they can run to make predictions or classifications without any knowledge of DL. Unlike other frameworks, with only a few lines of code, you can build a working web app. We will see an example of this in the hands-on part of this chapter (A use case for deploying a deep learning model as a web service – building a Streamlit application of the CNN model).

How to install Streamlit

Just like any other Python package, you can install Streamlit using the `pip` package manager. Just run `pip install streamlit` in your terminal or install Streamlit. The Streamlit version that will be used in this chapter is *version 1.12.2*.

To confirm whether Streamlit has been installed properly or not, just run `streamlit hello` in your terminal. This will launch an example web app in your browser shown as follows:

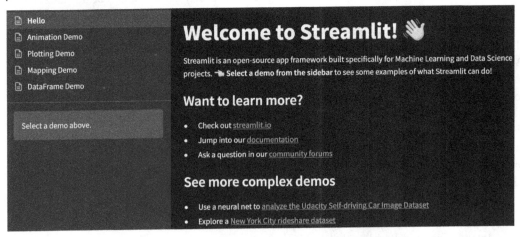

Figure 11.1 – Landing page of Streamlit using an example app

As soon as you type in that command, the preceding page should open automatically in your default browser (*Figure 11.1*). You can play with the options available on the side panel to understand more about the framework. We will go into the details in the hands-on section.

Hugging Face

Hugging Face is a very popular platform for the **machine learning** (**ML**) and DL community to host their models and datasets and create spaces. The platform provides tools that enable the community to build, train, and, most importantly, deploy their models. One of the capabilities of Hugging Face is **Spaces**, which has revolutionized how models can be deployed relatively easily with no costs. Spaces consist of easy-to-deploy features that use **Git** and provide friendly solutions for users to bring their apps to this platform. Currently, Spaces allows anyone to deploy Streamlit, Gradio, and HTML-based web applications, which fits in well for us since we will be deploying a Streamlit-based web app. In addition to simplicity in deployment, once deployed, the server runs fast and reliably with no downtime. If you are not importing your existing models, the platform provides seamless integration with the currently hosted models and datasets. We will see an example of how to deploy a Streamlit app using Spaces in the hands-on section (A use case for deploying a deep learning model as a web service – building a Streamlit application of the CNN model).

Introducing model deployment

When you start a DL project, your primary focus should be on data collection, data processing, and model development. The process of model development from data collection to model training always happens in an offline setting. However, wouldn't it be nice if the awesome model that you worked so hard on turned into something that other people can interact with? Even great models won't have much impact if they remain in the notebooks. Also, you wouldn't run your notebook every time new data comes in, right? So, how do you do it? The simple answer is model deployment, which is the process of integrating the model into an existing production environment to make appropriate business decisions based on data. It is the second-to-last stage of the DL life cycle before model monitoring and is the most important step for solving business challenges. A model that's deployed in production has several advantages:

- It will serve the needs of other users or researchers who can simply use the deployed model without them setting the environment to make the predictions

- The model will be scaled as per the requirements of the users

- It can perform real-time or batch predictions on live data

> **Note**
> Before the model gets deployed, you must make sure that the model is evaluated and tested properly and that it is fit to be in the production environment. The model must be tested for performance, efficiency, bugs, issues, and so on. Once you check all these boxes, the model is ready to be deployed.

Deploying models into production is challenging because of transitioning the model built on notebooks to the production environment, where the model can be run and used for prediction purposes. Because of this, most of the models don't get into production. The process requires a special set of skills and expertise and most data scientists and researchers do not know this. In organizations where there are large data science teams, the data scientists generally turn data into models and validate them; the MLEs are responsible for deploying the models (*Figure 11.2*):

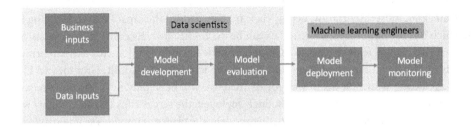

Figure 11.2 – Model deployment pre-steps

However, this may not be true for all data science teams since a small data science team will work on all aspects of model development, model validation, model deployment, and model monitoring. The process of model deployment varies greatly, depending on the system environment, the type of model, the availability of the existing resources for model deployment in the organization (for example, **Development and Operations (DevOps)** processes), and so on.

The key to model deployment is planning and that is the main reason why many models fail to be deployed in the production environment. This is because many organizations start DL projects without a deployment plan in place and they only think about it after model development. This will lead to delays in model deployment, which increases expenses for model deployment. There are a few key things you should consider during the planning stage, such as where you will store the data, what type of frameworks and tools you will use, how you will get feedback from deployed models, and so on.

Steps in model deployment

In general, there are seven key steps in model deployment as you are transitioning a model from research and development to production. Let's break down these key steps in model deployment that any genomic researcher or data scientist can use to deploy their models successfully:

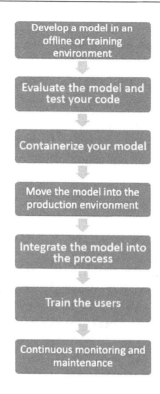

Figure 11.3 – Flowchart for the steps in model deployment

Let's look at each of these in detail:

1. **Develop a model in an offline or training environment**: Before you can deploy your model, you need to train your model in an offline setting or a training environment, as discussed in *Chapter 9, Building and Tuning Deep Learning Models*. Generally, multiple models are built as part of a DL project, but only one model gets deployed into production.

2. **Evaluate the model and test your code**: As you saw in *Chapter 9, Building and Tuning Deep Learning Models* after you built the model, you evaluated the model to ensure it was performing well on a test data based on evaluation metrics. This is very important because model deployment is a very involved and expensive process and only optimized models should be deployed to production.

3. **Containerize your model**: As briefly introduced previously, containerization allows models to be packaged reproducibly, so it's a very important tool for model deployment. Containerization has become very popular for DL model deployment because of its ease in model deployment and model scaling.

4. **Move the model into the production environment**: This is where the model is moved into an environment, where it has access to the hardware resources it needs to run (for example, Hugging Face) as well as the data source that it can draw data.

5. **Integrate the model into the process**: This can include, for example, making the model accessible from an end user's laptop using an API or integrating it into software currently being used by the end user.

6. **Train the users**: The people who will be using the model need to be trained in how to activate it, access the data, interpret the output, and so on.

7. **Continuous monitoring and maintenance**: Even though the model may work initially post-deployment, you shouldn't assume that the model will perform well in the long term. Continuously monitoring for model and concept drifts should be performed for the model to be effective for a long time.

Let's cover the different types of model deployment.

Types of model deployment

Models can be deployed across a wide range of environments, such as on-premise, the cloud, or hybrid. But before you deploy the model, it is important to understand what type of deployment method you want to use and where exactly your model fits in with the overall architecture of your application. In general, there are two main types of model deployment: batch prediction mode and online prediction mode.

Batch prediction

In this type of prediction, you run the model offline periodically on the new data, store the predictions in a database, and then return the results to the user if they query it. This type of prediction is done on problems for which there is no immediate need to generate predictions. The model predictions don't need to be run all the time and can run at a fixed schedule (once or twice a day). This is simple to implement and has low latency for the user. However, the predictions are not live, and if there are any issues with bugs, then the user will not know it. The main advantage of this method of model deployment is its ability to deploy complex models.

Online prediction

Many times, it is necessary to provide real-time interference, and, in this method, the results are generated in real time. A use case for online prediction would be, for example, a **Transcription Factor Binding Site** (**TFBS**) prediction, where the user inputs are expected to return corresponding predictions (yes or no) as seen in *Chapter 9, Building and Tuning Deep Learning Models*. Even though this method sounds exciting, this comes with some limitations, such as latency constraints. As the predictions are made in real-time, it is challenging to deploy complex models using this method.

Deploying models as services

In real-world projects, models are being used by other apps. Models are either integrated into apps through an REST API so that they can be accessed by the end user through HTTPS, or they are made directly available as an API. Deploying models through REST API is beyond the scope of this chapter and so we will skip that method here.

Model deployment through web services

Model deployment through a web application or service is the most preferred way to deploy models because it is easy for other researchers or data scientists to use without writing any code. In case if you are wondering what a web service is, a Web service is a server running on a computer device, listening for requests at a particular port over a network, serving web documents (HTML, JSON, XML, Images), and creating web applications services, which serve in solving specific domain problems over the web (www, internet, HTTP) (`https://en.wikipedia.org/wiki/Web_service`).In this type of deployment, you have a local client, which is either a web server or web app that the user has access to and which is running for them. It connects to a server where your model is deployed. Optionally, it connects to the database where it is pulling the data. Finally, the predictions are shown to the end user. There are two types of predictions in this type of deployment. Let's take a look.

Model deployment through model-in-service

In this type of deployment, your model is integrated into the server. This includes packaging your model and deploying it on the web server. In turn, the web server loads the model and uses that to make predictions. This method uses the existing infrastructure. However, this method has many drawbacks, such as the web service being written in a different language than your model. For example, your model may have been developed using Python and the web server might be using Node.js.

Model deployment through model-as-service

There is a modification of the model-in-service type where, instead of deploying the mode in the server, you will deploy it separately and both the client and web server can talk to that model directly. So, in essence, the model runs on its web server and the client interacts with this model and receives responses. This can be done using your Streamlit app, which we will see an example of soon. The main advantage of this method is its dependability, scalability, and flexibility.

Managed solutions

Whether you are leveraging REST APIs or web applications, you need to manage the dependencies that your model requires. The model predictions depend on code, model weights, and other dependencies and they all need to be present on the web server that you will be using to deploy your model. Managing dependencies is a hard problem to solve. You might simply copy the code and model weights (in the form of a pickle file) onto the web server, but other dependencies such as the versions of the libraries, packages, and so on are extremely hard to deal with. You can do so by implementing one of two different strategies.

ONNX

You can constrain the dependencies in the model by using a standard neural network format such as ONNX. ONNX is an opensource format that was built for ML models. It defines a common set of operators and a common file format so that DL developers can use models with a variety of frameworks, tools, runtimes, and compilers. You define the DNN in any language and can run it anywhere consistently. The second strategy is to use containers via Docker. Docker has become the standard method for managing dependencies due to it having several advantages, such as being lightweight, scalable, and able to containerize different parts of the dependency (for example, containerizing the code, web app, database, and so on).

Deployment tools

As you saw previously, there are a lot of steps in the model deployment process. Thankfully, there are managed tools to make the job of deploying a model to a server easier than ever before. They can take your model and turn it into an API, and you can interact with the API using HTTP requests or using gRPC to run inference on the server and still ensure that the dependencies are met and scaled. The following are some of the best model deployment tools out there for you to manage and scale your models:

- **TensorFlow Serving**: TensorFlow Serving is a high-performing and robust system for deploying models aimed at production environments. TensorFlow Serving makes the process of deploying the model easy with its out-of-the-box integration with TensorFlow models. It can also be extended to other types of models such as Keras, PyTorch, and so on. This deploys your model as an endpoint on a REST API where your model is currently being served.

- **MLflow**: Unlike enterprise tools such as Amazon SageMaker, MLflow is an open source tool for managing the model process and deploying your models. In addition to deployment, it can do other things, such as experimentation, reproducibility, and model registry. MLflow is compatible with different ML and DL libraries. One key point about MLflow is its ability to automatically track models.

- **Cortex**: Cortex is another open source tool for serving models, as well as monitoring your models. It has end-to-end workflows for model management operations.

- **Amazon SageMaker**: Amazon SageMaker is by far the most popular deployment tool. It is not just a tool but a powerful service provided by Amazon that gives data scientists and others the ability to develop, train, evaluate, deploy, and monitor models (both ML and DL). It makes the process of ML and DL super easy. As you know, the DL life cycle is fully iterative with complex steps and processes. Sometimes, it gets frustrating to perform all of these steps manually and it results in decreased efficiency and increased costs. Amazon SageMaker removes the painful manual steps in the DL life cycle process and makes the process super easy. The result is that the DL life cycle process is accelerated and deploying models is faster.

It is beyond the scope of this chapter to go into the details of the managed solutions We will use one of the open source managed solution with is Hugging Face which was introduced previously. In the next section, we will explore an example of building a simple Streamlit application for the model we trained in *Chapter 9, Building and Tuning Deep Learning Models*, using the Hugging Face platform.

A use case for deploying a DL model as a web service – building a Streamlit application of the CNN model

In this section, we will learn how to deploy a DL model as a web service for TFBS prediction. In this example, we use the Streamlit framework to wrap the CNN model that we built in *Chapter 9, Building and Tuning Deep Learning Models*. Once the app is deployed, through this web app, users will be able to upload a DNA sequence and see whether the sequence contains the TFBS or not

To create a web service, we need the following four steps:

1. The first step is to create a DL model, train it, and evaluate its performance.
2. In the second step, we need to persist or export the trained model.
3. In the third step, we will serve the persistent model through a Streamlit web application.
4. In the final step, we will be take the Streamlit application and deploying that on the Hugging Face platform.

Let's see all these steps in action.

Creating a DL model

We built a TFBS CNN model and evaluated its performance on test data in *Chapter 9, Building and Tuning Deep Learning Models*, so we will skip this step for now.

Exporting the DL model

Generally, the environments where the model is trained are very different from the deployment environment since the resource requirements vary a lot. Because of this reason, we will need to export the model from the training environment to the deployment environment. While training the model in *Chapter 9, Building and Tuning Deep Learning Models*, we used a callback in the Keras ModelCheckpoint to save the model to a file called `best_model.h5`, so we can also skip this step for now.

Creating the Streamlit-based application

This step, along with the next, are the most crucial for model deployment. Now that the model has been trained, evaluated, and exported to a file, we are ready to wrap the model in a web application and use that to make predictions on new data. Streamlit, as we discussed earlier, was chosen because of its simplicity in building web applications quickly. Since Streamlit is relatively new not many design options are currently available, unlike those in other frameworks such as Flask and Django. However, for the current web app, we only require basic features, which are fortunately available in Streamlit.

Let's look at the code for creating a simple Streamlit application of the CNN TFBS model that we built in *Chapter 9, Building and Tuning Deep Learning Models*:

1. First, we must create a helper script (`utils.py`) to prepare the input DNA sequence for our TFBS application. The script will do the following:

    I.   Receive the input DNA sequence and break it into 101 base pairs, starting from the first base pair.

    II. Create a dictionary, with keys being identifiers and values being the 101 base pair sequences.

    III. Take the 101 base pair sequences and convert them into a one-hot encoding matrix.

    IV. Return the dictionary keys and values and a one-hot encoding matrix:

```python
# Function for when you want to prepare DNA sequence
feature for ML applications
import numpy as np


# Function for when you want to prepare DNA sequence
feature for ML applications
def dnaseq_features(seq):
    start=0
    n_segs=101
    seq_name = 'seq'
    remaind = len(seq)%n_segs
    if(remaind != 0):
        last_id = len(seq) - remaind
    upd_seq = seq[start:last_id]
    dic_seq = {}
    for i in range(0,3):
        a = int(i*n_segs) ; b = int(i*n_segs)+n_segs
        identifier = f"{seq_name}_{a}:{b}"
        dic_seq[identifier] = upd_seq[a:b]
    lst_seq = dic_seq.values()
    index = list(dic_seq.keys())
    values = list(dic_seq.values())

    # One hot encode
    ii=-1
    for data in lst_seq:
```

```
        ii+=1
        abc = 'ACGT'
        char_to_int = dict((c, i) for i, c in
    enumerate(abc))
        int_enc = [char_to_int[char] for char in data]
        ohe = []
        for value in int_enc:
            base = [0 for _ in range(len(abc))]
            base[value] = 1
            ohe.append(base)
        np_mat = np.array(ohe)
        np_mat = np.expand_dims(np_mat,axis=0)

        if(ii != 0):
            matrix = np.concatenate([np_
    mat,matrix],axis=0)
        else:
            matrix = np_mat

    return matrix,index,values
```

2. Next, we will create a `requirements.txt` file that specifies the dependencies that our application requires. The content of that file will include numpy, `streamlit`, `keras`, and `tensorflow`.

3. Now, it's time to build a Streamlit web app. Unlike other web frameworks, Streamlit's code is very clean and with only a few lines of code, you will be able to build a Streamlit web app using only Python.

Let's understand the Python script that creates our Streamlit-based web app:

I.   First, we will export the Streamlit, NumPy, and Keras libraries. We will also import a script (`utils.py`) that will convert the input DNA sequence into a one-hot encoding matrix and get it ready to be fed during the prediction. We will use the Keras library to import the model that we previously saved in *Chapter 9, Building and Tuning Deep Learning Models*.

II. Next, we have a few lines of code that group the Streamlit and Python code. This block is used to specify the title and a caption for our app.

III. Then, we will create a horizontal line and a new container that has a section that inputs the DNA sequence using a function that we imported previously (`utils.py`).

IV.   Next, we add the horizontal line and then the section that launches the predictions using the input data. In this section, we have a run button to make predictions and display the results in a nice user-friendly manner.

```python
import streamlit as st
from keras.models import load_model
model = load_model('best_model.h5')
with st.container():
    st.title('Simple Model Serving Web App for TFBS prediction')
    st.caption('Get TFBS Predictions From The Latest Model.')
st.markdown("---")
with st.container():
    dna_seq = st.text_area("Input DNA sequence", 'ATAGAGAC...')
    dna_ohe_feat, ds_index, ds_val = dnaseq_features(seq=dna_seq)
    trigger = st.button('Make Prediction')
    if trigger:
        st.info("Loading the data for predictions")
        predicted_labels = model.predict(dna_ohe_feat)
        for i, j in zip(ds_val, predicted_labels):
            st.write(i)
            if np.argmax(j) == 1:
                st.success("TFBS found :thumbsup:")
            else:
                st.error('TFBS not found :thumbsdown:')
```

V.   Let's test this app with an example DNA sequence to predict if the user-provided DNA sequence contains any **TFBSs**. Starting Streamlit is very easy. We simply type the following code in our terminal:

```
$ streamlit run TFBS_prediction.py
```

You will see an output similar to the following in your terminal:

```
You can now view your Streamlit app in your browser.
Local URL: http://localhost:8501
Network URL: http://192.168.1.173:8501
```

Streamlit will also automatically launch the app using `localhost` on your default browser. The following screenshot (*Figure 11.4*) shows what our web app looks like at launch:

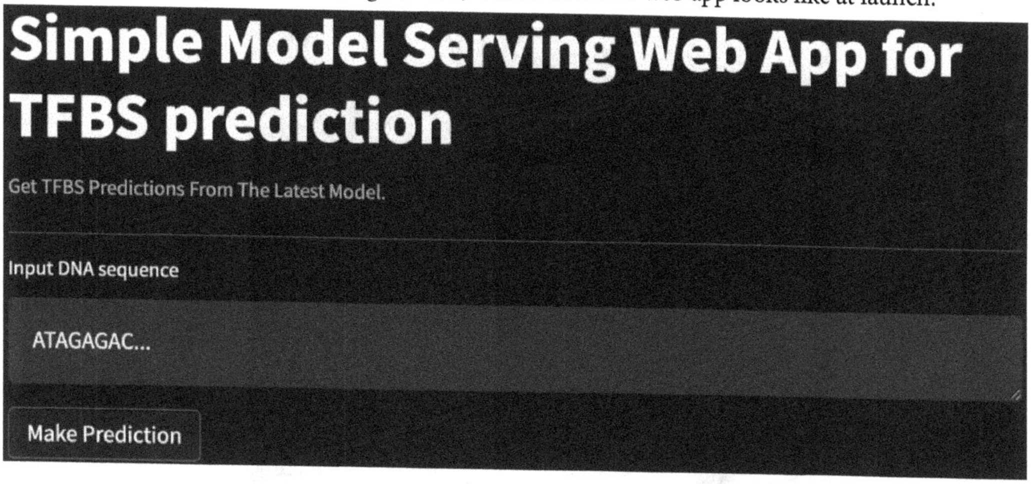

Figure 11.4 – The Streamlit app user interface for TFBS prediction

VI.   Let's copy and paste the following example DNA sequence into the **Input DNA sequence** section and click **Make Prediction** in the app:

```
ATGGCCGACCTCACTTCCCTTACGCAACATGGACGTTTTCAGTCACTTCCTGTGATT
TTGAGGTTATTCTGATCCAGGACTACTTGCATGACCAAAGCCAAGATGGCAGTTCCG
CCATGGAGTCATAGTTGCAGCTGACTCCAGGGCTACAGCGGGTGCTTACATTGCCTC
CCAGACGGTGAAGAAGGTGATAGAGATCAACCCATACCTGCTAGGCACCATGGCTGG
GGGCGCAGCGGATTGCAGCTTCTGGGAACGGCTGTTGGCTCGGCAATGTCGAATCTA
TGAGCTTCGAAATAAGGAACGCATCTCTGTAGCAGCTGCCTCCAAACTGCTTGCCAA
CATGGTGTATCAGTACAAAGGCATGGGGCTGTCCATGGGCACCATGATCTGTGGCTG
GGATAAGAGAGGCCCTGGCCTCTACTACGTGGACAGTGAAGGGAACCGGATTTCAGG
GGCCACCTTCTCTGTAGGTTCTGGCTCTGTGTATGCATATGGGGTCATGGATCGGGG
CTATTCCTATGACCTGGAAGTGGAGCAGGCCTATGATCTGGCCCGTCGAGCCATCTA
CCAAGCCACCTACAGAGATGCCTACTCAGGAGGTGCAGTCAACCTCTACCACGTGCG
GGAGGATGGCTGGATCCGAGTCTCCAGTGACAATGTGGCTGATCTACATGAGAAGTA
TAGTGGCTCTACCCCCTGAAAGAGGGTGGATGCAGCTGCTTGTGTTTCTTGGGGTGA
CTGTCATTGGTAATACGGACACAGTGACCCATCCTCCATCCTATTTATAGTGGAAGG
GCCTTCAATTGTATCAGTACTTTTTTTTAAGCTCTGGCACATTGACCTCTATGTGTT
ACCAGTCATTAATGAGCTGCTGCAGAGGTGACTATTTGTTTTACTTTCTTGGATGTT
AACATTACACTACTCACTACTCAATCTCAAAAAAAAAAAAAAAAAAA
```

Note

When you copy and paste this sequence in the app, make sure that there are no new line ('\n') characters in the sequence.

VII. After you copy and paste this sequence into the app (*Figure 11.5*), you will see the predictions that are being made by the app that correspond to the 101 bp for the sequence:

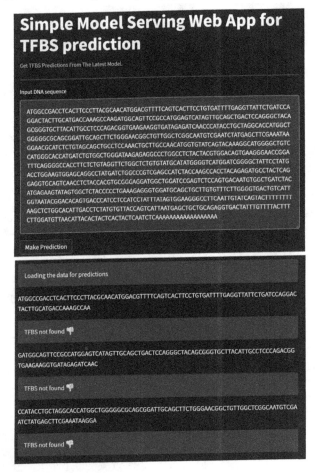

Figure 11.5 – Input and output from the app

As you can see, the app extracts the 101 bp from the sequence, runs the predictions, and returns the predictions along with the sequence. This is cool, isn't it?

Model deployment in Hugging Face Spaces

We have now reached the last step of the model deployment process, which is to take the current model, along with the web app that was tested locally, and move them to the production environment so that others can use them. Let's see how this can be done using Hugging Face Spaces.

1. First, create a new Space on Hugging Face by visiting their website (`https://huggingface.co/spaces`) and then clicking on **Create new Space**, as shown in *Figure 11.6*:

> **Note**
>
> Make sure you register on their website first before visiting this page.

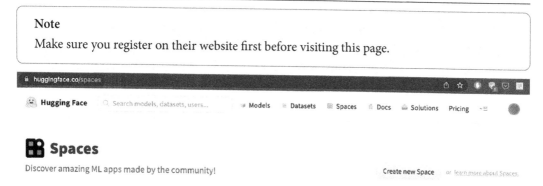

Spaces

Discover amazing ML apps made by the community!   Create new Space   or learn more about Spaces

Figure 11.6 – Creating an example Space on the Hugging Face website

2. Then, fill in the details of the Space. For this current app, here is what I have filled in (*Figure 11.7*):

Create a new Space

A Space is a special kind of repository that hosts application code for Machine Learning demos
Those applications can be written using Python libraries like **Streamlit** or **Gradio**

Owner   Space name

dl-genomics   ∨   /   predict-tfbs

License

License

Select the Space SDK

You can chose between Streamlit, Gradio and Static for your Space. Contact us if you need a custom solution.

| Streamlit | Gradio | Static |

○ Public
Anyone on the internet can see this space. Only you (personal space) or members of your
organization (organization space) can commit.

Private
Only you (personal space) or members of your organization (organization space) can see and
commit to this space.

Create space

Figure 11.7 – Creating an example Space on the Hugging Face website using the Streamlit option

3. Clicking the **Create space** button after carefully filling in the fields and selecting Streamlit will take you to the next step of the process.

4. The next step after creating the Space is to upload all of the Streamlit files to the Hugging Face GitHub repository. If you have used GitHub before, then the following section should be very familiar to you:

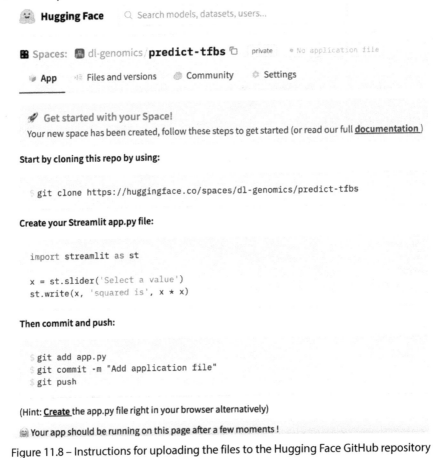

Figure 11.8 – Instructions for uploading the files to the Hugging Face GitHub repository

5. The most important step is to clone the repository locally using the link in the instructions. In this instance, you must run the following command:

```
$ git clone https://huggingface.co/spaces/dl-genomics/
predict-tfbs
```

> **Note**
>
> Please do not clone this repo as this space is setup for a demo. Please create a new space and follow the rest of the steps.

6. Next, you need to upload the three files (`app.py`, `utils.py`, and `requirements.txt`) that you have locally:

    I.   First, copy the three files (`app.py`, `utils.py`, and `requirements.txt`) into this GitHub repository.

    II.   Then, add and commit the files.

    III. Finally, push the changes to your GitHub repository.

After a few minutes of building the app, it will be readily available for you to use at `https://huggingface.co/spaces/dl-genomics/predict-tfbs`:

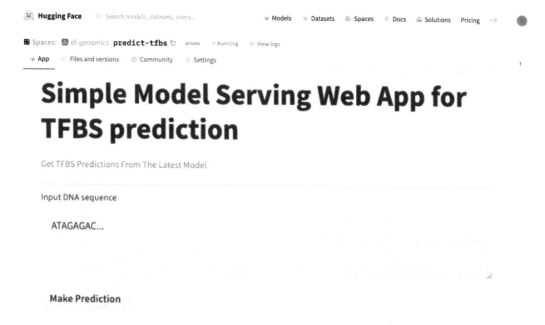

Figure 11.9 – The Streamlit app running on Hugging Face Spaces

This page should look familiar to you by now since this is the same page you saw when the app was running locally. Now, you can test your DNA sequence for the presence of TFBSs or not use the deployed app.

Depending on the settings, you can also share the link with anyone and they will be able to use this app to make predictions. Did you see how easy it is to take a model that is running locally to a production environment and share it with others?

Monitoring models using advanced tools

So far, you have built and evaluated the CNN TFBS prediction model (*Chapter 9, Building and Tuning Deep Learning Models*), interpreted it (*Chapter 10, Model Interpretability in Genomics*), and deployed it successfully (this chapter). You have even ensured that the model is working smoothly and correctly in a production environment. So, you might be thinking you are done, right? Not even close – but this is the beginning of a new journey. Just imagine what could go wrong after model deployment. Models can start to degrade post-deployment and consistently not perform the way they are expected to. Even though you have done everything right from model building to model deployment, things can go wrong after the model goes live in the production environment. Even after you have troubleshooted and tested a model thoroughly, things can still go wrong after model deployment.

Why monitor models?

Theoretically, once a model has been deployed, the model's performance should be consistent and robust to changes in the data coming from the real-world environment. However, it is quite common that a change in the environment leads to a drop in model performance and that the model may not be making the correct predictions. Model monitoring is the last phase of the DL life cycle and is where you check for inconsistencies in model predictions, even if the model has been trained, evaluated, and deployed properly. Model monitoring is done mainly to identify model underperformance or production issues so that we can intervene and iterate as quickly as we can to drive business value. This is important for models that are deployed in healthcare, financial, and other sensitive domains where model monitoring provides the detailed audit trails needed for compliance and risk management.

Reasons for model degradation

All models tend to degrade post-deployment. Why does this occur? Well, there are several reasons for models to underperform. They can be related to issues in the deployment environment or infrastructure, such as computing resources, memory, and so on. They can be related to the health of the model itself. Models are trained on data, so the predictive power of the model depends on the quality of the trained data. As time goes by, there will be shifts in user behavior or data changes, so the model's accuracy starts to degrade. These shifts can be instantaneous (for example, bugs in the pipeline), gradual (changes in user behavior), or periodic (some preferences are seasonal). These shifts can either have a lower impact or a huge impact in practice. In summary, there are two main contributing factors to the model's underperformance: data drift and concept drift.

Data drift

Here, changes in the underlying data expectations that your model is built on making the training data less relevant to the current situation (*Figure 11.10*). This may be caused because of issues in the deployment environment or when the data that is used to get predictions in the deployed environment is not clean, similar to the data cleaning process in the development environment. The data can also tend to change. This can include changes in the format, new categories, new fields, and so on:

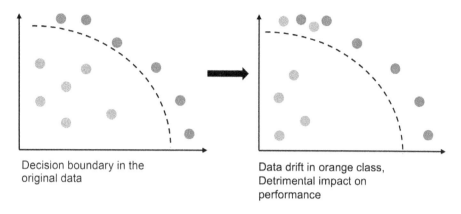

Decision boundary in the
original data

Data drift in orange class,
Detrimental impact on
performance

Figure 11.10 – Example of data drift

In summary, the training data is not representative of the actual data that is being used in the production environment. The result is less precise model predictions, as shown in *Figure 11.10*.

Concept drift

Unlike data drift, here, the data distribution doesn't change between the training data and production data. Instead, the underlying relationships between the inputs and the outputs of the model have changed, resulting in the model's predictions being less relevant and thereby changing the fundamental **concept** that the model needs to approximate, as shown in the following figure:

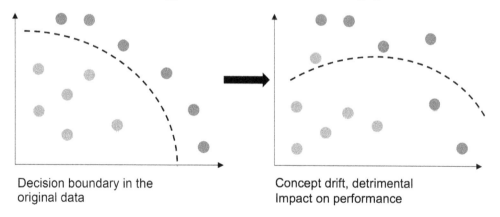

Decision boundary in the
original data

Concept drift, detrimental
Impact on performance

Figure 11.11 – Example of concept drift

As shown here, the model learned the decision boundary during training between the blue and orange classes. However, after model deployment, concept drift happened, and the relationships between the input and output in the data from the real world changed.

How to monitor DL models

Now that we know that monitoring models are important, let's look at how we can monitor models or detect these data and concept drifts. It all comes down to monitoring the model inputs and outputs in both the training and production data. You can monitor model metrics such as accuracy to detect when data drifts or concept drifts occur unexpectedly. In addition, there are other types of metrics, such as business metrics, model inputs, predictions, and system performance to monitor for DL models. These four types of metrics can be divided based on the level of information that they can provide or how easy they are to measure.

To detect data drift, we need to take the distributions of the input features and compare that to the production data, one feature at a time, and measure changes in the distribution using some metrics. To detect concept drift, since the relationship between the input and output changes, we should compare the conditional distribution of the truth against the predicted target for a given set of features. Since these calculations need to be done continuously as the production data is typically getting fed into the system regularly, it is recommended to automate this whole process as much as possible. We'll learn about the automated tools for model monitoring in the next section.

Advanced tools for model monitoring

The best way to monitor the model is to frequently evaluate the model's performance in the environment where it has been deployed and compare the predictions with the original outputs. Even though this approach sounds exciting, this is manual and error-prone. This should be an automated process that uses tools that can track model metrics and automatically alert you if there are any issues. Both enterprise and open source solutions exist for monitoring models, depending on the type of monitoring. They can be classified into three types of tools. Let's take a look.

System monitoring tools

These tools raise alarms and send notifications when things go wrong. Most cloud providers, including AWS, Azure, and GCP, have these types of tools. Some examples include Amazon CloudWatch, Datadog, and Honeycomb, which are mainly used to test traditional performance metrics.

Data quality tools

Some simple data statistics tools also exist for monitoring, such as **TensorFlow Data Validation** (**TFDV**). It has two great functions. The first function is generate_statistics_from_dataframe(), which can calculate simple statistics such as the mean, standard deviation, min, and max values, % of missing data, % of zero values, and so on. The second function is visualize_statistics(), which can be used to interactively inspect the data statistics. Other advanced tools such as Great Expectations, Anomalo, and Monte Carlo can test whether a particular window of data violates any rules or assumptions.

DL monitoring tools

This includes tools such as Arize, Fiddler, and Arthur's test models.

Addressing drifts

So far, we have investigated how to monitor models and how to detect drifts (data and concepts) if there are any in the deployed models. What if we do notice drifts in our system? The natural answer is that we retrain the data. However, the real question is how to retrain the model. The easiest way to retrain the model is to collect all the data after the drift and simply retrain the model's parameters based on that data. What if the drift was observed very recently, and this method doesn't help much because there is not much data to retrain the models? In practice, there is no one solution for this problem. This is usually problem- and domain-dependent. However, if the drift is significant, it might be necessary to optimize the hyperparameters for the new production data.

Summary

A successful application of DL for a genomics problem heavily relies not only on developing an accurate model but also on how to make the model impactful. Model deployment is the process of transitioning a trained model built on notebooks into a production environment where it is used for prediction, classification, clustering, and other purposes. Unlike model training, deploying models requires different skills that are not traditionally taught to data scientists and other genomic scientists because these skills, such as web app development, cloud computing, and working with APIs, are more software development skills. As the boundaries between data scientists and MLEs become blurred, knowledge of model deployment will take researchers a long way. In this chapter, you were introduced to a simple workflow for deploying the built model using some open source and easy-to-implement tools. These tools are easy to use and allow you to deploy a web app that can predict TFBS in a quick time.

Model monitoring should not be considered an optional step, and you should always be monitoring your models considering that the cost of wrong predictions can be huge because of a degraded model. Model monitoring generally starts after the model deployment phase, but it should be an integral part of the DL life cycle. It should not just be considered a post-deployment task – it should be done in both the training and evaluation phases of the DL life cycle, along with the post-deployment phase. The model tends to perform poorly due to the resources not being enough, the data quality being bad, and improper use of the application. Model monitoring ensures that the model is devoid of any performance issues and that predictions are correct. Not only does the model have to be monitored but also the entire infrastructure, such as software, resources, and so on. So far, we have looked at the different neural network architectures for genomics and understood how to build, interpret, deploy, and monitor DL models. In the final chapter, you will understand some of the challenges, pitfalls, and best practices of DL for genomics. See you in the next chapter!

12
Challenges, Pitfalls, and Best Practices for Deep Learning in Genomics

Deep learning (DL) is the branch of **machine learning** (ML) that encompasses **deep neural networks** (**DNNs**), with many artificial neurons arranged into several layers that mimic the human brain. Recently, DL algorithms have shown impressive results in several domains of life sciences and biotechnology. Even though the application of DL in genomics is relatively new, it has led to a fundamental understanding of biological and cellular processes in living systems. This has prompted many companies to leverage DL to solve important problems such as drug development, treatment of patients, and disease control, ultimately paving the way toward the promise of precision medicine in near future. Genomics is a data-rich discipline and is replete with complex datasets and often ill-understood. DL's success in genomics is largely attributed to its ability to perform knowledge extraction and pattern recognition from complex data automatically as opposed to domain scientists having to do it manually. In addition, the scale of current genomic data being produced, sophisticated algorithms, and availability of easy-to-use DL frameworks have propelled interest in the genomics community, resulting in the active application of DL for genomics research.

To make DL more accessible to genomic researchers, domain scientists, and others, they must have a proper understanding of the concepts of DL for genomics, assumptions behind DL models, and performance evaluations. This chapter will inform readers of the various challenges and the impact of various common pitfalls when applying cutting-edge DL methods to genomics. To help address these challenges and avoid pitfalls, this chapter will provide some guidelines, best practices, and solutions to help readers build end-to-end DL models and apply the same to genomic datasets. It is important to note that the list of challenges, pitfalls, and best practices included here is not exhaustive as the goal here is to introduce the basic ideas only.

With that in mind, here is the list of topics for this chapter:

- Deep learning challenges regarding genomics
- Common pitfalls for applying deep learning to genomics
- Best practices for applying deep learning to genomics

Deep learning challenges regarding genomics

The recent explosion of genomics data due to the advancements in **next-generation sequencing (NGS)** coupled with improvements in omic technologies (transcriptomics, proteomics, and metabolomics) has led to a greater understanding of the biological process of the living cell. Meanwhile, the remarkable success of DL based on DNN, has brought enormous improvements in **computer vision (CV)**, **natural language processing (NLP)**, and **machine translation**, and this has attracted the attention of genomics. The field of genomics quickly leveraged these specialized neural network architectures that can perform various tasks, such as binding site identification using CNNs, improving code optimization for improved protein translation through RNNs, unsupervised DL through autoencoders to predict gene expression, and so on. This is particularly exciting because genomics requires a data-driven and sophisticated solution to extract meaningful biological insights from these large and complex genomic datasets in an automated manner without the need to write rules by domain scientists. However, this comes with several challenges because of the complexity of genomics data. Let's review some of the potential challenges in applying DL for genomics.

Lack of flexible tools

Even though DL has shown tremendous promise in genomics, its adoption for addressing the current challenges in genomics by the genomics research community is slow. What could be the primary reason for this? Well, this is partially due to a lack of flexible DL models and tools to suit the new data. Even though those tools exist, it often requires significant effort to engineer the data before these tools can be used. In addition, other challenges include the inability to integrate multiple datasets or it not being suited for new neural network architectures.

Fewer biological samples

One of the primary reasons for DL's success in the fields of CV and NLP is that these models are routinely trained on huge data. However, this is not the case for genomics as it is still expensive to generate or it is experimentally not feasible to generate training data for that many samples in many biological contexts. For example, medical datasets are often relatively small, which increases the chance of spuriously correlated features in the datasets. Though it is reported that DL can be considered for datasets as low as 100 samples per class, it is generally recommended to have many samples for DL to be successful. A lot of data is deposited in public repositories such as SRA, GTEx, TCGA, gnomAD, UK biobank, 1K genomes, and so on. However, the challenge with these datasets is they have a lot of technical artifacts, and integrating those for mining insights through DNN is a bottleneck.

Computational resource requirements

Training DL models require significant computational resources, such as extensive computational infrastructure as well as patience to beat the state-of-the-art machine learning models or bioinformatics approaches. This is because of several millions of parameters and several hyperparameters that need to be tuned to make it right. So, this is a truly costly and time-consuming procedure. Though developing DL models in genomics rarely require this extensive training, as seen in *Chapter 9*, *Building and Tuning Deep Learning Models*, they still require computational resources that are beyond what traditional laptops and desktops can afford. With the availability of GPUs and cloud-based frameworks, the computational time required to build these models has come down significantly, but they do add extra costs to the genomic projects, which many companies and organizations cannot afford. In contrast, ML models rarely require specialized computational resources, so if building DL models don't add much that can already be done with ML, traditional approaches based on ML or bioinformatics methods may be the way to go for genomics.

Expertise in DL frameworks

The current challenges in using DL for genomics do not stop at requiring more data and computational resources. Building DL models often require more expertise than training ML models as training DL models correctly is nontrivial. Traditionally, genomic researchers heavily rely on bioinformatics tools for data analysis, which are mostly straightforward. This is not the case for building DL models in genomics. Even with the availability of DL frameworks such as TensorFlow, PyTorch, and Keras, which allow users to quickly prototype and build models easily with little programming experience, this is still daunting for new genomic researchers. More recently, **automated ML** (**AutoML**) eliminated the need for manual intervention at each step of the DL life cycle as it helps with feature engineering, feature selection, model selection, and other steps in the DL life cycle automatically. For example, AutoML tools such as H2O, DataRobot, AutoKeras, and so on can build, test, and optimize the DL models automatically and can allow users to only involve, when necessary, with a few lines of code. This way, they can reduce the expertise needed to build and use DL models. They are already capable of abstracting away much of the programming required to build "standard" DL tasks. However, those tools are expensive, less flexible, and require a substantial learning curve.

Lack of high-quality labeled data

Like any other real-world data, genomic data also suffers from poor quality, and even if it is available, not all the data might be labeled. This is a problem when using supervised models. Despite this limitation, DL has the potential to apply specific research questions and problems in genomics. If quality labeled data (for classification tasks) is available, DL can be successfully applied to genomics.

Lack of model interpretability

One of the main challenges of DL today is the "black-box" nature of its models. This is generally a problem in domains such as biological and clinical, where they expect transparency in models that are being used. Because of this, many DL models don't end up in production in these disciplines.

Unlike simpler ML models such as regression and classification, which can be interpreted very easily, it is challenging to interpret, for instance, what node and layer are important for model performance in DL. Because DL uses deep network architectures, nonlinear activation functions, and several parameters, which enables them to achieve a very high prediction accuracy, this comes at a huge cost, which, in turn, makes these models less interpretable. The lack of interpretability might not be a huge problem if used for not-so-critical projects and fields. However, for those models that are being used in clinical and healthcare settings, the lack of model interpretability will become an important issue. This is because the model trustworthiness of DL models is one of the most important requirements for model deployment in these fields.

Common pitfalls for applying deep learning to genomics

The genomics field has undergone a big data revolution with the advent of NGS, which has allowed researchers to take molecular measurements such as gene expression at a genomic scale. This technological advancement has led to a greater understanding of cellular and biological processes and has shown promise for treating many uncurable diseases in clinical settings. As the amount and complexity of genomic data increased, researchers started to leverage DL to extract useful biological information and build predictive models. This has led to many DL tools being used for a wide variety of genomic analysis tasks, such as processing raw data, integrating heterogeneous datasets, predictive modeling, and so on. To prevent low model performance when applying DL for genomics data, there are common pitfalls that one should be aware of. Let's discuss the common pitfalls that you might face when trying to apply DL to genomic tasks and how to avoid them.

Confounding

Confounding is where an unmeasured or artefactual variable ("confounder") creates or masks association with an outcome. This occurs because the confounder induces dependence between features and outcomes. Because of this, models can pick up these correlations as part of the learning process, even though they are not related to scientific questions at hand. The result will be misguided predictions and wrong conclusions. To avoid this common pitfall, you can use unsupervised learning and **exploratory data analysis (EDA)** to identify these biases early in these datasets before leveraging a supervised DL model.

Data leakage

Data leakage is also a common pitfall in DL modeling. When the data from the test set leaks into the training data, this is known as **data leakage**. This results in highly accurate models and results in overfitting when tested on real-world datasets. After splitting the data into training, validation, and holdout (test) datasets, you must ensure the test data is kept separate until the final step of model development (model evaluation) to prevent data leakage.

Imbalanced data

This is a very common pitfall in genomics for supervised learning tasks. In supervised learning, each data point is associated with a label. It is known as balanced data if the samples have evenly distributed classes or unbalanced otherwise. Let's explore this concept a bit further with a few examples. Let's consider a dataset representing cancer labels – control and cancer. In general, for these kinds of datasets, there are very few samples that have cancer labels associated with them. Datasets are rarely perfectly balanced, and many problems in genomics show extreme imbalance. Class imbalance is the default state in genomics. Many problems in genomics are imbalanced and researchers routinely employ statistical methods to ensure the positive samples are real. In the **transcription factor binding site** (**TFBS**) example, which we covered in *Chapter 9, Building and Tuning Deep Learning Models*, there are a lot of 101 bp genomic windows that have no TFBSs (negative) compared to the positive sequences that contain TFBSs. Another classical example is the disease prediction risk in *Chapter 3, Machine Learning Methods for Genomic Applications*; only a certain percentage of patients have the disease compared to the whole dataset.

The main drawback of this data is the model tries to over-learn from the majority class and under-learn from the minority class. This is fine for those cases where false negatives are not a problem, but it is in the case of disease prediction, where the minority class is of primary interest and the cost of false negatives comes at a premium. Another example is predicting the side effects of a drug. Supervised regression tasks also suffer from data imbalance. Several strategies can mitigate imbalanced data, such as putting prior probability distributions on the classes, oversampling the minority class or undersampling the majority class, weighting examples, and so on. Let's review some of them now:

- We have already seen how GANs can be used to create artificial data to address the issues of data imbalance.

- Another thing to watch out for when using imbalanced data is the evaluation metric, which is used to evaluate the imbalanced data; for example, accuracy. As an example, a model that has been trained on 990 negative samples and 10 positive samples will give 99% accuracy by predicting the negative class. This means the model that predicts the negative class will have high accuracy. Hence, other evaluation metrics such as the ROC-AUC curve and AUC metric should be used instead as both quantify recall and precision, as you learned in previous chapters.

- Another important thing to remember is that oversampling and undersampling methods should only be performed on the training data. If you do undersampling or oversampling on the whole data and then split the data into train and test datasets, then the test split distribution may not represent the distribution of real-world data and hence undersampling or oversampling should be avoided before splitting.

In summary, in genomics, a common pitfall is using genomics data without addressing the data imbalance. Even though this pitfall focuses on the classification problem, similar problems exist in regression tasks.

Improper model comparisons

Comparing the performance of simpler models, such as ML versus advanced methods such as DL models, should be done carefully. Comparing baseline conventional ML models trained with default settings with the DL-trained models that went through rigorous model training and hyperparameter optimization should be done carefully. Such comparisons can lead to false conclusions about model efficacy. Unequally tuning the hyperparameters in the simpler base models compared to carefully tuning the hyperparameters in DL-trained models can skew model evaluation. It is recommended to perform a similar level of model tuning for the simpler models before making a claim about which model is better compared to the other. A performance comparison is only informative when both models are equally well optimized.

Best practices for applying deep learning to genomics

So far, we have seen several challenges with DL for genomics and common pitfalls and how to avoid them. To make the best use of DL for genomics, let's look at a collection of some best practices to follow while leveraging DL for genomics.

Understand the problem and know your data better

As seen in the DL life cycle, the most important aspect of DL is understanding the business goal or scientific question that you are trying to solve and then framing the business goal into a DL problem. Having a proper understanding of the scientific question (business goal) and a clear analysis plan (framing the business goal into a DL problem) are key to the success of DL projects in genomics. You should not even start working on DL without defining the goals of the project. For instance, would you step into the lab without thinking about what you plan to do that day? No. Right? Some of the key questions that you should be asking to include, does the data exist for this problem, does this data answer the biological question, how do I source the data, how much data can I get, and what metadata exists? Once you have reviewed these goals, then you should gather the data and understand the study protocol early in the process: everything starts with data.

Frequently, the data, such as TFBS data, is sourced from public databases or publications, as we discussed previously. However, this data might not be appropriate for your needs, and you need to transform the data to make it fit your problem. The data may be unstructured, metadata may be missing, the sample size may be smaller than expected, and so on. These problems will limit the data's usability and ability to build better DL models. Documenting the data sources and their dates properly will ensure reproducibility. Once the data has been obtained, it is important to learn what and why the data has been collected before beginning the actual analysis. This can be done using metadata, which comes with the data.

Always know your data very well so that you can understand the context and peculiarities. This will help you avoid the pitfalls described earlier. The more time you spend understanding the data, the less time you spend addressing the pitfalls. Genomics data is very complex and to make the best use of DL to solve genomic problems, it's better to spend some time understanding the data.

A simple model for a simple problem

Just because you know how to build DL models doesn't mean you should throw DL at every problem in genomics. The state-of-the-art methods that rely heavily on bioinformatics and computational biology guided by rules work for many problems in genomics. Those methods are peer-reviewed, well-supported, and easy to apply. DL models, on the other hand, are not only hard to train but also hard to interpret, deploy, and, most importantly, reproduce. In addition, not all genomics projects involve NGS data. Many times, they contain less data, and in those circumstances, non-DL methods are suited to that problem. More data does not mean DL can outperform other conventional methods. Sometimes, simpler models achieve performance the same or better than DL with fewer resources. Well-tuned DL models generally tend to outperform classical ML and other methods but many times, simpler models such as ML models or rule-based methods perform very well, especially when the data noise increases.

Establish a baseline for your model

Once you frame the business question into a DL problem and before even applying a DL model to a genomics problem, it is best to implement a simpler model that doesn't require tuning lots of hyperparameters. Here, the simple model represents a baseline for your DL model. Examples include ML models such as logistic regression, random forests, naïve Bayes, and so on. These simple models can set the baseline expectations and difficulty of the problem in the study. During the model evaluation process, in addition to the model evaluation metrics, you can compare your DL model with a simple model. If your DL model is performing poorly in comparison with the simple well-tuned model, that indicates there is no added advantage of deploying this DL model. Instead, either continue with the simple model or improve the performance of the DL model.

> **Note**
> One pitfall to avoid is that you should carefully compare two different models, as discussed previously.

As a best practice, it is recommended to use the baseline model while using the same software framework as DL for consistency in the pipeline that was used. There are instances when you can combine these simple models with DL models to build a hybrid model that is more accurate than any of the individual models.

In addition to selecting a non-DL model to establish a baseline for comparison purposes, starting a simple neural network architecture using a few layers and nodes per layer will help you understand the model's complexity. Depending on the model's performance, the model's complexity can be increased. This will reduce the time it takes to build a good model as you will be wasting a lot of time and computational resources building a highly complex model from the beginning. Sometimes, simple decisions such as choosing an optimization algorithm can have a significant impact on the final model.

Ensure reproducibility

Reproducibility should be the norm in any scientific discipline and DL should not be an exception. It's important to ensure reproducibility is practiced in DL because, unlike a typical bioinformatics project, DL model training involves a lot of steps, from data collection to model monitoring. Furthermore, to obtain a good model, it is important to optimize a lot of parameters, known as **hyperparameters**. Often, you need to repeat this process multiple times. All of these will impact reproducibility. The code that was used, the random seed that was set, the hyperparameters that were used to tune the model, and the results must be collected to ensure reproducibility. Some of the popular tools for reproducibility include Git (`https://git-scm.com/`) for version control of your code, Docker (`https://www.docker.com/`) for packaging code and dependencies, DVC (`https://dvc.org/`) for data version control and GitHub Actions for **continuous integration and continuous development (CI/CD)** (`https://github.com/features/actions`). Ensure reproducible tools are used during model training, model optimization, and model deployment so that anyone else can follow and reproduce the same on their end.

Using pre-existing models for genomics

As much as possible, leverage existing trained models that have been built for genomics. Many of the models, such as DL models for TF prediction, TFBS prediction, or microRNA identification, can be used to make predictions without you having to build the models from scratch. Many genomic tools use prebuilt models, so it is highly advisable to use those. Along the same lines, if the current genomic tools or DL models are either not suited for the current problem or not performing well, then it is advisable to use the transfer learning method that you learned previously. This will reduce the burden on the process of DL model development and still generate good results. As a best practice, it would be good to revalidate the results that are obtained from these pre-existing models to make sure that the data and distributions match with the data that we have at hand. Frameworks such as WeightWatcher (`https://github.com/CalculatedContent/WeightWatcher`) can be used to ensure that the pre/trained model that you are interested to leverage for your problems are not over-trained or over-parameterized without or without training data.

Do not reinvent the rule

Technically, this should be the first best practice for DL. Do not reinvent the rule. The current state of DL is very advanced due to the availability of neural network architectures that are suited to many tasks in many disciplines, including genomics. Rather than trying to design a new neural network architecture, use the existing architectures that have shown promising results in genomics and other disciplines.

Tune hyperparameters automatically

Hyperparameter tuning is an integral part of training DL models. Very rarely one can build a model that performs well without tuning hyperparameters because of the complexity of the data, network architecture, and so on. Many times, very simple neural network architectures such as one hidden layer with multiple nodes and nonlinear activation functions such as ReLu or slightly complex networks that have multiple layers that can approximate continuous functions may not suit the complex problems. Hence, deeper architectures with multiple hidden layers and multiple hidden units per layer are used. This, in turn, results in many more parameters and hyperparameters to tune, which pose additional challenges when training DL models. In addition to the number of layers and number of nodes, hyperparameters involve choosing an optimization algorithm, loss function, learning rate, nonlinear activation function, training batch size, regularization penalty, dropout, batch normalization, and so on. Because of the complexity of these hyperparameters, it is impossible to evaluate which neural network methods are well suited for a particular task manually. This requires extensive and significant effort toward hyperparameter optimization. Thankfully, tools such as KerasTuner that you have learned in *Chapter 9, Building and Tuning Deep Learning Models* and Ray Tune (`https://github.com/ray-project/ray/tree/master/python/ray/tune`) support automatically tuning hyperparameters without manually optimizing the same. As a best practice, it is highly recommended to use these automated solutions for hyperparameter optimizations in DL.

Focus on feature engineering

Feature engineering is complex but it is one of the most critical components of building a good model. Traditionally, for genomic data, the input data is a DNA sequence that gets transformed into a one-hot encoding. But there are instances where you need to engineer the features that will be provided along with the sequence data. Understanding how the data is collected can help you decide how to engineer features. Since genomics is a highly domain-specific field, it is always a good practice to involve domain experts in discussions during feature engineering. During the feature selection step, select a feature that generalizes well and remove the features that aren't contributing to the model's performance. Good feature engineering comes with a better understanding of the business problem and the domain.

Normalize the data

Like any other data, genomic data also contains features that have a different range of values. So, it is a good practice to normalize or scale the data before building a model. Instead of using the entire data, use the statistics from the training data to gain information about the features and normalize or scale them or handle missing data.

Always perform model interpretation

Often, genomic researchers build a model to get model insights rather than predictions. Either way, it is important to perform model interpretation, as described in *Chapter 10, Model Interpretability in Genomics*. Model interpretation helps us understand the model's behavior regarding why some decisions have been made by the model (either correct predictions or false predictions). Researchers very rarely perform model interpretation because either they are more interested in the predictions, or they don't care why and how the model has made those decisions as long as it made the right decisions.

However, this is one of the most critical components of building models today, especially for genomic and clinical applications. Use tools that interpret these "black-box" DL models that were successfully used in other domains such as CV and NLP. Many of them are applied in genomics with much success.

Avoid overfitting

Building models that generalize well on the test data is the goal of DL. However, it is not always easy to build models that perform well on the test data as it is with the training data. Overfitting occurs when the model performs well on the training data but, when subjected to the test data, which was not exposed during model training, it does not perform well. It's like us memorizing exam materials and, when confronted with completely new questions in the exam, we struggle to answer questions that we have not studied. This is exacerbated in DL because of the large number of parameters and hyperparameters used during model training.

The first way to start addressing the overfitting problem is to detect it. You can do this by splitting the entire data into training, validation, and testing datasets. These three partitions can help optimize models by iterating between model training using the training set and hyperparameter optimization using the validation dataset. Validation data can be used to look at the overfitting of the model as you can compare the trained model to the validation data. For example, you can plot the training loss at each epoch versus the validation loss. If the training loss is the same as the validation loss for the epochs trained, then we know the model has been trained and it did not overfit.

> **Note**
> Test data should only be used for final model evaluation and should not be used for hyperparameter tuning.

When you detect overfitting, you can use methods such as data splitting, regularization, data augmentation, dropout, early stopping, building simpler models, and gaining awareness of biological non-dependence to address them.

Summary

DL has shown great promise in genomics and these methods now match or exceed the current state-of-the-art methods in a diverse array of tasks and disciplines in life sciences and biotechnology. Given this rapid rise in its applicability across broad research areas to understand the complexities of biological systems, its adoption is still low. This can be attributed to several factors, the major among them being the complexity of genomic data. In this final chapter of this book, we have seen some of the challenges and common pitfalls associated with applying DL in genomics, which reduces the effectiveness of these DL methods. Addressing these challenges and pitfalls can be hard because of the complexity of the DL methods. Often, the mistakes are very subtle, and you didn't know that you are making them. To avoid making these simple mistakes, you must have a deep understanding of the concepts surrounding genomics and DL. One major piece of advice is that if the genomics scientists and researchers can become familiar with the data and the algorithm, and then conduct robust follow-up analysis, this will pave the way to ensure the effective use of DL in genomics while striving for high model interpretability and high model performance.

Despite these challenges, DL has been dominant over competing traditional technologies such as ML approaches in many applications. We are in an era where we are witnessing DL models achieving human-level performance in many genomics tasks, all thanks to the current advancements in sequencing technology, algorithms, and next-generation hardware. When used effectively, DL can be very powerful, as seen in several applications of genomics, and cooperation between domain scientists and DL algorithms addresses many of the current challenges in genomics and can help you achieve incredible results. DL for genomics is there to stay and we should expect some incredible achievements in the future.

This chapter, as well as the rest of this book, intended to introduce you to the concepts surrounding DL for genomics with the goal that these serve as building blocks toward building successful models that can address some of the challenging problems in life sciences and genomics disciplines. I hope that you will take these concepts and knowledge and use them either in your research or in building production models for your projects.

Good luck in your DL journey in genomics!

Index

A

activation function 60
  binary step function 60
  leaky ReLU 62
  ReLu 61
  sigmoid 62
  softmax 63, 64
  tanh 63
activator protein (AP) 175
advanced tools, model monitoring
  data quality tools 222
  DL monitoring tools 223
  monitoring 222
  system monitoring tools 222
AI, in genomics market
  reference link 6
Amazon SageMaker 210
Amazon Web Services (AWS) 17
Analysis using Denoising Autoencoders
    for Gene Expression (ADAGE)
  reference link 130
anomaly detection 119
  novelty detection 120
  outlier detection 119, 120
area under the curve (AUC) 172
artificial neural networks (ANNs) 4

artificial neuron (AN) 57
association 121
autoencoder applications, for
    predicting gene expression
  ADAGE 130
  gene expression clustering, boosting 131
  hierarchical organization, of yeast
    transcriptomic machinery 131
autoencoders 73, 74, 121
  architecture 124
  bottleneck 125
  convolutional autoencoders 127, 128
  decoder 125
  deep autoencoders 127
  encoder 125
  gene expression 130
  gene expression use case 131
  image compression 125
  loss function 125
  properties 122
  regularization 125
  regularized autoencoders 128
  Single-cell RNA sequencing
    (scRNA-seq) 126, 127
  vanilla autoencoder 127
  working 122-124
automated ML (AutoML) 227

B

backpropagation 64, 65
backpropagation through time (BPTT) 102
balanced data 229
bases 13
Bayes algorithm 139
bias 60, 68
bidirectional RNN 103
biological neuron 57
Biopython 7, 18
  FASTA 20
  GenBank 21
  installation, verifying 12
  installing 10
  installing, pip used 11
  Python installation, verifying 10
  SeqIO object 20
  seq object 18, 19
  SeqRecord object 19
  sequences, working with 18
  using, for genomic data analysis 18
black-box model interpretability 185, 186
bottleneck 125
business value, unlocking from
      model interpretability 187
  business decisions 187-189
  profitability 189
  trust, building 189

C

ChIP-seq experiment
  TFBS predictions via 111
ChIP-sequencing (ChIP-seq) 38
classification metrics or performance
      statistics, model evaluation
  accuracy 170, 171

  Precision 171, 172
  Recall 171, 172
clustering 37
CNN architecture 84
  convolutional layer 84, 85
  fully connected layer 86, 87
  input layer 84
  output layer 87
  pooling layer 86
CNN for coexpression (CNNC) 93
coding sequence (CDS) 35
computer vision (CV) 137, 226
considerations, for algorithm
      implementation
  memory requirements 161
  model interpretability 161
  training time 161
continuous integration and continuous
      development (CI/CD) 232
contractive autoencoders 129
convolutional autoencoders 127, 128
convolutional neural networks
      (CNNs) 70, 71, 82, 83
  applications, in genomics 90
  for genomics 89, 90
  history 83
Cortex 210
cross-validation 163

D

DanQ 108
  reference link 108
data collection 157, 158
data leakage 228
data partitioning 161, 162
  cross-validation 163
  dataset, training 162

Group K-Fold cross-validation 163

holdout dataset 162

K-Fold cross-validation 163

random partitioning 162

stratified K-Fold cross-validation 163

stratified partitioning 162

validation dataset 162

data preprocessing

data augmentation 159

data cleaning 158

data transformation 159

data processing 157

data transformation

data processing 16

quality control and cleaning 16

data wrangling

data preprocessing 158, 159

decoder 125

deep autoencoders 127

deep autoregressive models 137

DeepBind 90

DeepChrome 92

DeepInsight 91

deep learning (DL) 4, 51, 56

DragoNN 80

Kipoi 80

life cycle 154-156

neural network definition 56

workflow, for genomics 74, 75

deep learning, applying to genomics

best practices 230-234

deep learning challenges

computational resource requirements 227

expertise in DL frameworks 227

fewer biological samples 226

lack of flexible tools 226

lack of high-quality labeled data 227

lack of model interpretability 227, 228

deep learning libraries

Keras 79

PyTorch 78

TensorFlow 78

deep learning model

creating 211

deploying, as web service 211

exporting 211

monitoring 222

DeepNano 107

reference link 106

deep neural network (DNN) 58, 121, 159

activation function 60

anatomy 56

application, in genomics 76

architectures 69

backpropagation 64

bias 60

for genomics 74

forward propagation 64

gene regulatory networks (GRNs) 77

gradient descent 65

hidden layer 58

input layer 58

key concepts 59

loss function 64

output layer 58, 59

protein structure predictions 77

regularization 65

regulatory genomics 77

single-cell RNA sequencing 77

transfer function 60

weights 60

DeepTarget 108, 109

reference link 108

DeepVariant 92

denoising autoencoders 129

denoising autoencoders, used for predicting gene expression from TCGA pan-cancer RNA-Seq data 131

data collection 131

data preprocessing 131

model training 132-135

deoxyribonucleoside triphosphates (dNTPs) 13

deployment tools

Amazon SageMaker 210

Cortex 210

MLflow 210

TensorFlow Serving 210

dimensionality reduction technique (DRT) 121

dinucleotide content

calculating 25

discriminative models

versus generative models 138, 139

DNA methylation 66

DNN application

gene expression prediction 76

SNP prediction 76

DNN architectures

autoencoders 73, 74

convolutional neural networks (CNNs) 69

feed-forward neural networks (FNNs) 69

graph neural networks (GNNs) 69

recurrent neural networks (RNNs) 69

DragoNN 80

URL 80

drifts

addressing 223

dropouts 126

E

encoder 125

Encyclopedia of DNA Elements (ENCODE) 33

engineered features

versus learned features 159

exabytes (EB) 33

Explanation Summary (ExSum) 194

exploratory data analysis (EDA) 15, 38, 228

extract, transform, and load (ETL) 15, 158

F

False Positive Rate (FPR) 172

feature engineering 159, 233

feature variation 121

feed-forward neural networks (FNNs) 70, 138

feed-forward NNs (FNNs) 138

forget gate 104

forward propagation 64

fully connected (FC) layer 70

G

GANs applications 147

crease data, analysis 148, 149

DNA generation 149

for augmenting population-scale genomics data 150

gated recurrent unit (GRU) 104, 105

GC content

calculating 24

GenBank 33

gene expression 130

Gene Expression Omnibus (GEO) 41

Generative Adversarial Networks (GANs) 138-140

components 140

discriminator, training 141, 142

generator, training 142, 143

model improvement 146, 147

working 140, 141

generative models

versus discriminative models 138, 139

generator model 139

genome sequencing 12

anger sequencing, of nucleic acids 13

genome-wide association studies (GWAS) 146

genomic big data 33, 34

genomic data analysis 14

Biopython, using 18

cloud computing, for genomics data analysis 17

data collection 16

data transformation 16

dinucleotide content 25-27

exploratory data analysis 16

GC content, calculating 24

ML modeling 27-29

modeling 17

motif 29, 30

nucleotide content, calculating 24, 25

steps 15

visualization and reporting 17

use case 21-24

genomics

CNNs, applications 90

convolutional neural networks (CNNs) 89, 90

deep learning challenges 226

deep neural network (DNN), using for 74

DNN application 76

machine learning (ML), need for 5

ML, challenges 51

pre-existing models, using 232

recurrent neural network (RNN), applications 106

recurrent neural network (RNN), use cases 106

genomics datasets

working with, challenges 143, 144

Git 205

global surrogate method 192, 193

GoLan 107

gradient descent 65

graph neural networks (GNNs) 72

Group K-Fold cross-validation 163

H

hidden Markov models (HMMs) 139

Hugging Face 205

reference link 216

Hugging Face Spaces

model deployment 216-219

Human Genome Project 13

hyperparameter tuning 164, 232

Bayesian optimization 165

grid search 164, 165

libraries 166

libraries, KerasTuner 166-168

model training 181

random search 165

I

image compression 125

ImageNet Large Scale Visual Recognition Challenge (ILSVRC) 83

individual conditional

expectation (ICE) 191

input gate 104

J

JUND TF
  binding site location, predicting 174

K

Keras 79
KerasTuner 166-168
kernels 84
key performance indicators (KPIs) 155
K-Fold cross-validation 163
Kipoi
  URL 80
k-nearest neighbors (knn) 145

L

learned features
  versus engineered features 159
learning problem 34
life sciences
  machine learning for genomics 6
Local Interpretable Model-agnostic
    Explanations (LIME) 193
Logistic regression 40, 48
log-loss 173
long short-term memory (LSTM) 103, 105
loss function 125

M

machine learning (ML) 4, 5
  challenges, in genomics 51
  for genomics 38

  in biotechnology 6
  in life sciences 6
  need, for genomics 5
machine learning (ML), in genomics
  data collection, to preprocessing 38
  feature extraction and selection 39
  model deployment 40
  model evaluation 39
  model interpretability 40
  model monitoring 40
  model training 39
  train-test data splitting 39
  workflow 38
machine learning software
  exploring 6
machine translation 226
managed solutions 209
  deployment tools 210
  ONNX 210
Matplotlib 12, 32
mean absolute error (MAE) 36
mean squared error (MSE) 39, 118
median absolute deviation (MAD) 131
median absolute percent error (MAPE) 174
MLflow 210
ML libraries 32
  scikit-learn 33
ML use cases 40
  data collection 41, 42
  data preprocessing 42, 43
  data splitting 47, 48
  data transformation 45, 46
  EDA 43-45
  model evaluation 48-51
  model training 48
model
  tuning 163

model degradation
concept drift 221
data drift 220
reasons 220
model deployment 205
advantages 205
as services 209
batch prediction 208
managed solutions 209
model-as-service 209
online prediction 208
pre-steps 206
steps 206, 207
through model-in-service 209
through web services 209
types 208
model development
appropriate algorithm, selecting 161
steps 160
model evaluation 169
classification metrics or
    performance statistics 169
performance visualization 172
regression metrics 173
model interpretability 184
business value, unlocking from 187
methods 189
use cases 194
model interpretability methods
ExSum 194
global surrogate 192, 193
individual conditional
    expectation (ICE) 191
partial dependence plot 190
permuted feature importance (PFI) 191, 192
saliency map 194
Shapley value 193

models
degrading, reasons 220
monitoring, need for 220
monitoring, with advanced tools 220
model training
data partitioning 161
multidimensional scaling (MDS) 37
multi-layer perceptrons (MLPs) 69

N

nanopore sequencing 14
**National Center for Biotechnology
    Information (NCBI) 33**
natural language processing (NLP) 66, 226
**neural machine translation
    model (NMT) 108**
neural network (NN) 138, 185
definition 56
working, example 66-69
neural network architecture, creating 197
model evaluation 198
model interpretation 199, 200
model training 197, 198
next-generation sequencing (NGS) 13, 226
evolution 14
second-generation DNA sequencing
    technologies 14
third-generation DNA sequencing
    technologies 14
non-linearity 67
**normalized empirical probability
    distribution function (NEPDF) 93**
novelty detection 120
nucleotide 12
content, calculating 24, 25

O

outlier detection 119, 120
output gate 104
Oxford Nanopore Technology (ONT) 106

P

Pandas 32
parameter sweeping 164
partial dependence plot (PDP) 190
permuted feature importance (PFI) 191, 192
pip
  used, for Biopython installation 11
pitfalls, for applying deep learning
    to genomics 228
  confounding 228
  data leakage 229
  improper model comparisons 230
PLoS Biol
  reference link 5
polymerase chain reaction (PCR) 14
positional weight matrix (PWF) 5
position frequency matrix (PFM) 29
positioning weight matrix (PWM) 71
precision-recall (PR) curve 173
predictive modeling 17
principal component analysis
    (PCA) 121, 147, 158
ProLan 107, 108
ProLanGo
  reference link 107
Python
  download link 11
Python packages
  Matplotlib 32
  Pandas 32
  seaborn 32

Python programming language 6
PyTorch 78
  model and data 79
  model deployment 79

R

R2 174
receiver operating characteristic
    (ROC) 172, 173
Rectified Linear Unit (ReLu) 61
recurrent neural network (RNN) 96-99, 138
  applications, in genomics 106
  backpropagation 102
  understanding, through TFBS
    predictions 110
  use cases, in genomics 106
  working 99-101
recurrent neural network (RNN), types 105
  many-to-many 106
  many-to-one 106
  one-to-many 105
  one-to-one 105
regression metrics 173
  R2 174
  RMSE 173
regularization 65
  data augmentation 66
  dropout 66
  Elastic Net 66
  Lasso 66
  ridge 66
regularized autoencoders 128
  contractive autoencoders 129
  denoising autoencoders 129
  sparse autoencoder 128, 129
ReLU activation function 68
research and development (R&D) 187

reset gate 105

Ribo-sequencing (Ribo-seq) 38

RNA sequencing (RNA-seq) 77

RNN architectures 102

  bidirectional RNN 103

  gated recurrent unit (GRU) 103, 105

  long short-term memory (LSTM) 103, 105

root mean squared error (RMSE) 169, 173

S

saliency map 194

scikit-learn 7

seaborn 32

sequence record 19

sequencing by synthesis (SBS) 14

SHapley Additive ExPlanations (SHAP) 40

Shapley value 193

Single-cell RNA sequencing
  (scRNA-seq) 126, 127

single-molecule sequencing
  real-time (SMRT) 14

single nucleotide polymorphisms
  (SNPs) 35, 76

singular value decomposition (SVD) 37

Spaces 205

sparse autoencoders 128, 129

stochastic gradient descent (SGD) 164

stratified K-Fold cross-validation 163

Streamlit

  installing 204, 205

Streamlit-based application

  building, of CNN model 211

  creating 211-216

structural zeros 126

Structured Query Language (SQL) 158

style transfer 137

supervised learning (SL) 138, 162

supervised ML 34, 35

  types 35

  working with 36

support vector machines (SVM) 92

synapse 57

synthetic data, for genomics 145

  techniques 145, 146

Synthetic Minority Oversampling
  TEchnique (SMOTE) 145

T

Telomere to Telomere (T2T) effort 76

TensorFlow 78

TensorFlow Data Validation (TFDV) 222

TensorFlow Serving 210

tensors 78

TFBS prediction problem

  data collection 175

  data preprocessing 176

  data, processing 175

  data wrangling 175

  feature engineering 176-178

  framing, in terms of DL 175

  model training 179, 180

TFBS predictions, via
  ChIP-seq experiment 111

  data collection 111

  data preprocessing 112

  data splitting 112

  model training 112-115

The Cancer Genome Atlas (TCGA) , 33

Torch 78

Transcription Factor Binding
  Site (TFBS) 208, 229

Transcription Factor Binding Site (TFBS) predictions

recurrent neural network (RNN), understanding through 109, 110

Transcription factors (TF) 109, 188

transfer function 60

transfer learning (TL) 88, 121

True Positive Rate (TPR) 172

Truncated backpropagation through time (TBPTT) 102

U

Universal Approximation Theorem 56

unsupervised DL 118

anomaly detection 119

association 121

clustering 118

unsupervised learning (UL) 169

unsupervised ML methods 37

clustering 37

dimensionality reduction 37

types 37

update gate 105

use cases, model interpretability

data collection 195

feature extraction 195

Neurol Network architecture, creating 197

target labels 196

train-test split 196

V

vanilla autoencoder 127

visualization libraries, Python 7

W

weights 60

whole-exome sequencing (WES) 38

whole-genome sequencing (WGS) 38

whole-transcriptome sequencing (WTS) 38

Packt.com

Subscribe to our online digital library for full access to over 7,000 books and videos, as well as industry leading tools to help you plan your personal development and advance your career. For more information, please visit our website.

Why subscribe?

- Spend less time learning and more time coding with practical eBooks and Videos from over 4,000 industry professionals

- Improve your learning with Skill Plans built especially for you

- Get a free eBook or video every month

- Fully searchable for easy access to vital information

- Copy and paste, print, and bookmark content

Did you know that Packt offers eBook versions of every book published, with PDF and ePub files available? You can upgrade to the eBook version at packt.com and as a print book customer, you are entitled to a discount on the eBook copy. Get in touch with us at customercare@packtpub.com for more details.

At www.packt.com, you can also read a collection of free technical articles, sign up for a range of free newsletters, and receive exclusive discounts and offers on Packt books and eBooks.

Other Books You May Enjoy

If you enjoyed this book, you may be interested in these other books by Packt:

Practical Deep Learning at Scale with MLflow

Yong Liu

ISBN: 9781803241333

- Understand MLOps and deep learning life cycle development
- Track deep learning models, code, data, parameters, and metrics
- Build, deploy, and run deep learning model pipelines anywhere
- Run hyperparameter optimization at scale to tune deep learning models
- Build production-grade multi-step deep learning inference pipelines
- Implement scalable deep learning explainability as a service
- Deploy deep learning batch and streaming inference services
- Ship practical NLP solutions from experimentation to production

Machine Learning on Kubernetes

Faisal Masood, Ross Brigoli

ISBN: 9781803241807

- Understand the different stages of a machine learning project
- Use open source software to build a machine learning platform on Kubernetes
- Implement a complete ML project using the machine learning platform presented in this book
- Improve on your organization's collaborative journey toward machine learning
- Discover how to use the platform as a data engineer, ML engineer, or data scientist
- Find out how to apply machine learning to solve real business problems

Packt is searching for authors like you

If you're interested in becoming an author for Packt, please visit `authors.packtpub.com` and apply today. We have worked with thousands of developers and tech professionals, just like you, to help them share their insight with the global tech community. You can make a general application, apply for a specific hot topic that we are recruiting an author for, or submit your own idea.

Share Your Thoughts

Now you've finished *Deep Learning for Genomics*, we'd love to hear your thoughts! Scan the QR code below to go straight to the Amazon review page for this book and share your feedback or leave a review on the site that you purchased it from.

https://packt.link/r/1-804-61544-7

Your review is important to us and the tech community and will help us make sure we're delivering excellent quality content.

Download a free PDF copy of this book

Thanks for purchasing this book!

Do you like to read on the go but are unable to carry your print books everywhere?

Is your eBook purchase not compatible with the device of your choice?

Don't worry, now with every Packt book you get a DRM-free PDF version of that book at no cost.

Read anywhere, any place, on any device. Search, copy, and paste code from your favorite technical books directly into your application.

The perks don't stop there, you can get exclusive access to discounts, newsletters, and great free content in your inbox daily

Follow these simple steps to get the benefits:

1. Scan the QR code or visit the link below

https://packt.link/free-ebook/9781804615447

2. Submit your proof of purchase
3. That's it! We'll send your free PDF and other benefits to your email directly

www.ingramcontent.com/pod-product-compliance
Lightning Source LLC
Chambersburg PA
CBHW080633060326
40690CB00021B/4922